AF595465

Additively Manufactured Coatings

Additively Manufactured Coatings

Editors

Pradeep Menezes
Pankaj Kumar

MDPI • Basel • Beijing • Wuhan • Barcelona • Belgrade • Manchester • Tokyo • Cluj • Tianjin

Editors
Pradeep Menezes
Department of Mechanical Engineering, University of Nevada Reno
USA

Pankaj Kumar
Department of Mechanical Engineering, The University of New Mexico
USA

Editorial Office
MDPI
St. Alban-Anlage 66
4052 Basel, Switzerland

This is a reprint of articles from the Special Issue published online in the open access journal *Coatings* (ISSN 2079-6412) (available at: https://www.mdpi.com/journal/coatings/special_issues/3D_Print_Coat).

For citation purposes, cite each article independently as indicated on the article page online and as indicated below:

LastName, A.A.; LastName, B.B.; LastName, C.C. Article Title. *Journal Name* **Year**, *Volume Number*, Page Range.

ISBN 978-3-0365-2488-7 (Hbk)
ISBN 978-3-0365-2489-4 (PDF)

Contents

About the Editors

Pradeep Menezes is an Associate Professor in the Mechanical Engineering Department at the University of Nevada, Reno. Before joining the University, he worked as an Adjunct Assistant Professor at the University of Wisconsin-Milwaukee (UWM) and post-doctoral research associate at the University of Pittsburgh.

His productive research career has produced over 130 peer-reviewed journal publications (over 5500 citations, h-index—37), 30 book chapters, and 4 books. He currently serves as a reviewer on over 60 prestigious journals and as an editorial board member of 10 journals. In addition, he has participated in many national and international conferences in roles such as conference paper reviewer, conference review committee member, conference technical committee member, and session chairman. He has peer-reviewed multiple books, book chapters, grants, and master's and doctoral theses. He has also prepared technical newsletters and tribology web notes for the American Society of Mechanical Engineer's (ASME) tribology division.

Pankaj Kumar is an Assistant Professor in the Mechanical Engineering department at the University of New Mexico (UNM), Albuquerque. He obtained his doctoral degree from the University of Utah, Salt Lake City, USA. Following doctoral work, he explored the physical, mechanical, and electrochemical behavior of additive manufactured materials during his postdoc research at the University of Utah. Then, he joined the University of Nevada, Reno as a research assistant professor in the Chemical and Materials Engineering department, and continued exploring research in the additive manufacturing area.

Dr. Kumar's research is in the broader areas of advanced manufacturing and materials design for novel structural and functional applications. His research includes material concepts with specific emphasis on advanced manufacturing, materials processing, and physical and mechanical metallurgy. His research also includes improving understanding of the micro-mechanisms of materials damage such as fatigue, creep, etc

Editorial

Additively Manufactured Coatings

Pankaj Kumar [1,*] and Pradeep L. Menezes [2]

1 Department of Mechanical Engineering, University of New Mexico, Albuquerque, NM 87131, USA
2 Department of Mechanical Engineering, University of Nevada, Reno, NV 89557, USA; pmenezes@unr.edu
* Correspondence: pankaj@unm.edu

We are pleased to publish a Special Issue on "Additively Manufactured Coatings" that is intended to provide peer-reviewed articles in the fascinating field of coatings, particularly in the area of additive manufacturing technology.

Functional coatings are of particular importance in the protection of the substrate from chemical and mechanical damage in aggressive environments. They are widely used as cost-effective methods to protect the substrate from wear, corrosion, erosion, tribocorrosion, high temperature, and high pressure in extreme environmental conditions. These are primarily manufactured through metal/ceramic powder deposition in a subsequent layer-by-layer fashion on the substrate materials. In all cases, the functional coatings need to be reliable for the intended application. The emerging additive manufacturing techniques can be utilized to develop high-performance functional coatings. These methods provide geometrical precision, flexibility in geometrical complexity, customization of the coating layers, and reduce the raw material waste, keeping the manufacturing cost low while addressing many of the technical barriers of conventional coating methods. With the rapid development of cutting-edge value-added technologies in the aerospace, nuclear, military, space, and energy industries, additive manufacturing techniques will be major advantages. Novel functional coatings and additive manufacturing techniques will be critical to value-added components in the future development of technologies. The objective of this Special Issue is to include the recent advances in the development of functional coatings utilizing additive manufacturing techniques. This issue provides the latest information and knowledge on the recent developments of coatings utilizing the additive manufactuing techniques that will benefit the professionals from both industries and research and educational organizations with discipline including but not limited to mechanical, manufacturing, materials, aerospace, and chemical. Early career professionals and students will greatly benefit from the Special Issue on advanced coating technologies.

The editors of this issue are from academia while closely working with the industries. From both the industrial and research perspective, the benefits and limitations of additive manufacturing coating technologies are covered in six chapters of this issue. The contributors to this issue are from both academia and industries and possess good experience in coating technologies.

By utilizing additive manufacturing techniques, a sound coating has been produced. Both the solid-state and fusion-based additive manufacturing techniques can be utilized to produce a sound quality coating for various metals and alloys. For example, the cold spray (CS) based technique can produce high-quality titanium (Ti) coating for the magnesium (Mg) alloy substrate to increase its wear and corrosion resistance [1]. Many different coatings, such as Ti, and aluminum (Al), have been developed on AZ31B alloy using the CS technique. The study showed that Ti coating significantly improved the hardness, wear, and corrosion characteristics (in 3.5% wt NaCl solution) of AZ31B compared to the Al coating. The fusion-based AM technique, such as laser powder bed fusion (L-PBF), has been shown as a potential technique to develop the corrosion and tribocorrosion resistant coating on the meal or alloy substrate [2]. In this Special Issue, not only the potential use of the 3D technique for coating is covered, but the basic mechanisms of coating formation are

Citation: Kumar, P.; Menezes, P.L. Additively Manufactured Coatings. *Coatings* **2021**, *11*, 609. https://doi.org/10.3390/coatings11050609

Received: 15 March 2021
Accepted: 17 May 2021
Published: 20 May 2021

also included. Oyinbo et al. [3] studied the basic mechanism of layer formation in the CS technique using molecular dynamics while considering ductile materials as examples. They showed that due to the jetting phenomena at the particle/substrate interface, materials deform. This can cause a compressive wave formation of ~2.8 GPa at a particle speed of 1000 m/s in the impacted zone. The high-pressure wave is propagating into the substrate and particle from the point of impact leading to the deformation of particle and substrate and heat generation resulting in layer formation. Therefore, the formation of metal layers in CS is given by the new microstructure formation at the particle/substrate interface. Depending on the processing condition, the plasticity at the interfaces will change; so will the coating quality.

In one of the chapters, the quality of the coating in terms of the microstructure and the process parameters using full factorial design in the Laser-based technique is studied [4]. By manipulating the process parameters, the microstructure and, therefore, the wear characteristics of the composite substrate can be tailored. In addition, the thermal stress cycle in the Laser-based cladding process is studied for the nickel coating on steel using a theoretical approach [5]. It is shown that by using the stress analysis, the process parameters can be controlled to achieve the optimum coating conditions [5]. The evolution of the microstructure during the interlayer formation of Ni-based coating is studied in Laser-based technique [6]. He et al. [6] indicated that interlayer cooling has a significant impact on the microstructure and impurity of the interlayer. They showed that the hardness of the interlayer cooling specimens is greater than that of the non-interlayer-cooled specimens.

In the end, we thank all the authors for their valuable contributions. We are confident that you will find this Special Issue of interest to refer to for your learning needs in your coating technology research and applications.

Funding: This research received no funding.

Institutional Review Board Statement: Not applicable.

Informed Consent Statement: Not applicable.

Conflicts of Interest: The authors declare no conflict of interest.

References

1. Daroonparvar, M.; Kasar, A.K.; Farooq Khan, M.U.L.; Menezes, P.; Kay, C.M.; Misra, M.; Gupta, R.K. Improvement of Wear, Pitting Corrosion Resistance and Repassivation Ability of Mg-Based Alloys Using High Pressure Cold Sprayed (HPCS) Commercially Pure-Titanium Coatings. *Coatings* **2021**, *11*, 57. [CrossRef]
2. Siddaiah, A.; Kasar, A.; Kumar, P.; Akram, J.; Misra, M.; Menezes, P.L. Tribocorrosion Behavior of Inconel 718 Fabricated by Laser Powder Bed Fusion-Based Additive Manufacturing. *Coatings* **2021**, *11*, 195. [CrossRef]
3. Temitope Oyinbo, S.; Jen, T.-C. Molecular Dynamics Simulation of Dislocation Plasticity Mechanism of Nanoscale Ductile Materials in the Cold Gas Dynamic Spray Process. *Coatings* **2020**, *10*, 1079. [CrossRef]
4. Huang, X.; Liu, C.; Zhang, H.; Chen, C.; Lian, G.; Jiang, J.; Feng, M.; Zhou, M. Microstructure Control and Friction Behavior Prediction of Laser Cladding Ni35A+TiC Composite Coatings. *Coatings* **2020**, *10*, 774. [CrossRef]
5. Yao, F.; Fang, L. Thermal Stress Cycle Simulation in Laser Cladding Process of Ni-Based Coating on H13 Steel. *Coatings* **2021**, *11*, 203. [CrossRef]
6. He, Y.; Liu, Y.; Yang, J.; Xie, F.; Huang, W.; Zhu, Z.; Cheng, J.; Shi, J. Effect of Interlayer Cooling on the Preparation of Ni-Based Coatings on Ductile Iron. *Coatings* **2020**, *10*, 544. [CrossRef]

Article

Thermal Stress Cycle Simulation in Laser Cladding Process of Ni-Based Coating on H13 Steel

Fangping Yao [1,*] and Lijin Fang [2]

[1] School of Mechanical Engineering and Automation, Northeastern University, Shenyang 110004, China
[2] Faculty of Robot Science and Engineering, Northeastern University, Shenyang 110004, China; ljfang@mail.neu.edu.cn
* Correspondence: yyaofangping@163.com

Abstract: In order to improve the work efficiency and save resources in the process of laser cladding on the H13 steel surface, based on COMSOL, by combining computer simulation and experiment, a plane continuous heat source model was used to simulate and analyze the temperature and stress field. The optimal power and scanning speed were obtained. It is found in the simulation process that the thermal sampling points stress increases with the increase of laser power and scanning speed. Because of the existence of solid–liquid phase variation in the laser cladding process, there are two peaks in the maximum thermal stress cycle curve of the sample points located in the molten pool, and the starting and ending time of each sample point's peak value is basically the same. When the sample point is outside the molten pool, because the metal at the corresponding location is not melted, so there is no obvious peak value in the thermal stress cycle curve. With the increase of cladding layer depth corresponding to each sample point, the variation range of the two alternating thermal stress peaks increases first and then decreases, while the duration increases. According to the peak value of alternating thermal stress at the sampling point, the molten pool depth can be predicted. The residual stress analysis of the cladding layer is carried out according to the analysis results of temperature field and stress field. Through the actual cladding experiment, it is found that the depth of molten pool in the simulation results is basically consistent with the experimental results. All simulation results are verified through actual cladding experiments.

Keywords: laser cladding; H13 steel; thermal stress cycle; numerical simulation; unstable alternating thermal stress; residual stress

Citation: Yao, F.; Fang, L. Thermal Stress Cycle Simulation in Laser Cladding Process of Ni-Based Coating on H13 Steel. *Coatings* **2021**, *11*, 203. https://doi.org/10.3390/coatings11020203

Academic Editors: Pradeep Menezes and Pankaj Kumar

Received: 2 November 2020
Accepted: 4 February 2021
Published: 10 February 2021

1. Introduction

H13 steel is widely used as a kind of hot die steel. It can be used for manufacturing various hot extrusion and forging dies [1,2] because of its high thermal strength, hardness, abrasion resistance, toughness, good heat resistance, and fatigue performance. But the surface performance will be significantly reduced if the H13 steel mold is exposed to high temperature for a long time [3–5]. Laser cladding technology can effectively improve the surface performance of the metal, but the process of cladding is affected by many factors. If only a single experimental study is carried out, the work efficiency will be low and resources will be wasted [6]. By combining computer simulation and experiment, the research cycle can be greatly reduced and the efficiency can be improved [7].

In the past two decades, scholars domestic and abroad have established numerous laser cladding simulation models and conducted relevant studies [8]. Kong et al. [9] adopted the enthalpy-pore method to deal with the phase transition phenomenon in the cladding process, adopted the horizontal set method to track the movement of the molten pool. The forced convection and thermal radiation on the surface of the cladding layer were incorporated into the simulation model, while the heat transfer process in the laser multilayer cladding process of H13 section steel was studied. Liu Hao et al. [10] developed a simulation model of laser cladding temperature field for simultaneous powder delivery.

The model processed the additive effect in the entire cladding process through the life and death element method, the shielding effect of the powder flow on the high-energy laser beam and the heat transfer of the laser beam in the powder conveying process were considered by calculating the shading rate [11–15].

Kar and Mazumder [16] proposed a one-dimensional conduction model to determine the alloy composition and cooling process. Hoadley and Rappaz [17] developed a two-dimensional model to calculate the steady-state temperature in the laser cladding process, which gave a quasi-steady-state numerical model of the substrate temperature field. The calculation takes into account the changes in the liquid molten pool, the gas–liquid for the shape and position of the free surface, in order to simplify the model, it is considered that the substrate melts very little and uses the line energy form of the laser, so that the approximate linear relationship between the laser power, the scanning speed, and the thickness of the repair layer is obtained. Han et al. [18] solved the two-dimensional fluid and energy equations, predicted the temperature distribution and geometry of the molten pool during the laser cladding process. Cho and Pirch [19] published a three-dimensional steady-state finite element model of coaxial powder feeding. A self-consistent method is used to calculate the temperature field and coating shape. The obtained temperature gradient and cooling rate are used to predict the coating solidification structure. Jendrzejewski [20] discussed the influence of the preheating temperature on the temperature field and stress field of the repair layer, and adopted a linear approximation to its temperature characteristics. After preheating, the thermal stress value of the substrate repair layer reduced from 1800 MPa to 900 MPa, and got a crack-free repair layer. Toyserkani [21] et al. developed a three-dimensional transient finite element model of coaxial powder feeding. The coating is a multilayer structure, its width and height are determined by the area of the previous layer and laser powder. The model ignores the effect of surface tension and gravity on the shape of the coating. He [22,23] et al. studied the three-dimensional numerical model of the molten pool temperature and fluid flow during the laser cladding process of H13 steel, and used the level set method to simulate the molten pool. Kovacevic [24] and others at the Laser Aided Manufacturing Center of Southern Medical University USA used ANSYS to establish a finite element model. The law of temperature distribution and cooling rate in the molten pool under positive and negative defocusing conditions was studied. Compared with the Gaussian laser beam, it is found that the hollow laser beam (defocus) can effectively prevent the center of the molten pool from overheating, but the model does not consider the powder input impact when it reaches the molten pool. Parisa Farahmand studied the single multi-layer temperature field, strain stress field, and pool size evolution rule for laser cladding [25]. Gao W calculated the temperature variation, cooling rate, and solidification rate at the solid/liquid interface, but he did not consider the effect of Marangoni convection on the weld pool [26]. The Indian Institute of Technology and Monash University used H13 tool steel as the matrix to perform a single-pass laser cladding of CPM9V powder. It was found that the surface hardness of the cladding layer under single-layer and multi-layer conditions are the same. Researchers from Harazmi University in Tehran, Iran and Kaye Nasir Tutsi University of Technology in Tehran, Iran, used ABAQUS to simulate the laser cladding process of WC powder on the surface of Inconel 718. Analyzing the temperature field and residual stress distribution, it is concluded that the laser power and scanning rate have a greater influence on the dilution rate and residual stress of the cladding layer. The increase of input energy increases the residual stress of the cladding layer and the number of cracks decreases [27].

In this study, a planar continuous heat source model was used to conduct numerical simulation on the single-pass laser cladding process of H13 steel based on COMSOL software, the optimal process parameter scheme was determined, the thermal stress and thermal cycle curves were drawn and analyzed, which are used to study the influence of thermal stress cycle on the cladding layer, the residual stress of the cladding layer is also simulated.

2. Simulation Model

2.1. The Establishment of Finite Element Model

The matrix is H13 steel. The sample is a cuboid with the size of 60 mm × 60 mm × 10 mm. The cladding layer is added to the center of the matrix with a length of 60 mm, its height and width are determined by the powder feed amount and laser diameter. In COMSOL, Figure 1 shows the grid division.

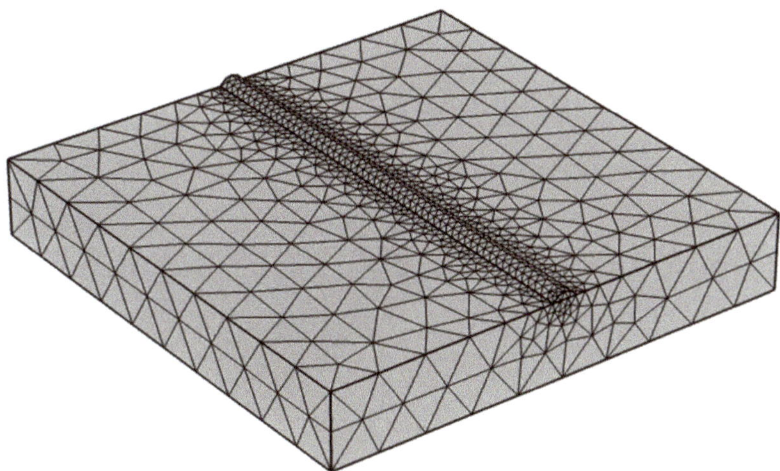

Figure 1. Mesh division of finite element model.

The establishment of heat source model will affect the accuracy of laser cladding numerical simulation results. Both plane heat source and body heat source are suitable for laser cladding. The plane heat source model is generally divided into pulse heat source and continuous heat source. Compared with other heat source models, planar continuous heat source is used more. In this study, the planar continuous heat source model shown in Figure 2 is selected. The distribution function is:

$$q(x,y,t) = \frac{Q}{\pi R_0^2} e^{-\frac{2[(x-x_0)^2+(y-v_0t)^2]}{R_0^2}} \tag{1}$$

where $q(x, y, t)$ is the heat flow at the (x, y) position of time t, Q is the laser power, R_0 is the laser beam radius, X_0 is the position of the laser center in x direction; v_0 is the laser cladding velocity.

The matrix used in this experiment is H13 steel, the cladding material is Ni60 alloy powder. The matrix and powder composition are listed in Tables 1 and 2. According to the chemical composition of the matrix and powder, the corresponding materials from COMSOL database were selected, and their physical properties provided by the software were used.

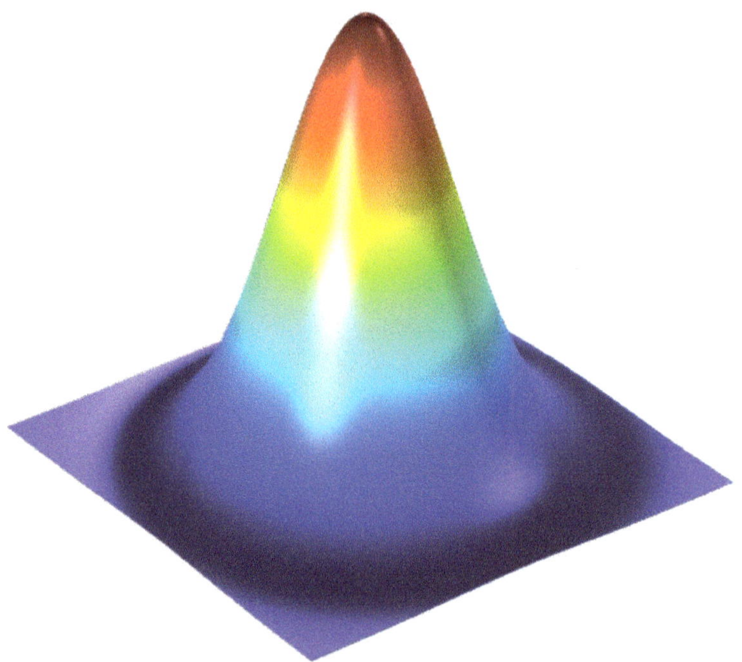

Figure 2. Planar continuous heat source model.

Table 1. Chemical composition of H13 steel.

Element	C	Si	Mn	Cr	Mo	V	P	S	Fe
Percentage content/wt%	0.38	0.92	0.28	5	1.2	0.95	0.02	0.03	Bal.

Table 2. Chemical composition of Ni60 alloy powder.

Element	C	Cr	Si	Fe	B	Ni
Percentage content/wt%	0.8	16	4.0	15.0 Ma	3.2	Bal.

2.2. *Simulation Parameters and Sample Points Distribution*

The main parameters of laser cladding include laser power, scanning speed, spot diameter, and powder feeding rate, among which the spot diameter and powder feeding rate mainly affect the height and width of cladding layer, but have little influence on the temperature field distribution. Through simulation calculation, it is found that the temperature field cloud map of different spot diameters and powder feeding rates is basically the same, which is not necessary to be analyzed, but mainly to analyze the influence of laser power and scanning speed on the laser cladding temperature field. The simulation was carried out by changing the laser power and scanning speed. The laser power was 600 W, 800 W, 1000 W, 1200 W, and 1400 W, the scanning speed was 2 mm/s, 3 mm/s, and 4 mm/s.

In order to work out the optimal experimental parameters and analyze the thermal stress cycle of the cladding layer, probes were added to 11 sample points in the cladding layer to record the changes of temperature and thermal stress in real time. Figure 3 shows the locations of the sample points. The cross section is selected in the middle part with stable cladding to ensure the representativeness and accuracy of sample points, avoiding excessive errors as well.

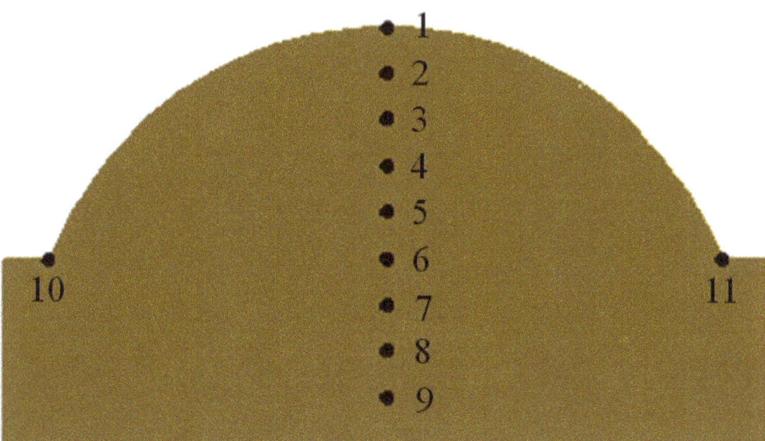

Figure 3. Sample points locations and number.

2.3. COMSOL Solution

In this simulation, the temperature field and thermal stress field of the laser cladding material were simulated. A planar continuous heat source shown in Figure 4 was used to simulate the laser heat source; the laser heat source is directly loaded onto the cladding surface, while the laser defocus is adjusted by the heat source radius. The required physical fields module can be added in COMSOL according to the requirements.

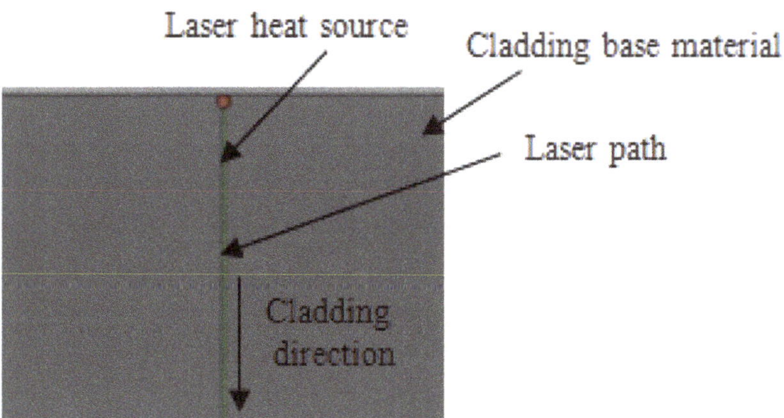

Figure 4. Laser heat source loading model.

The temperature field simulation needs to be calculated in "solid heat transfer" physical field, while the solid heat transfer and thermal stress field simulation physical fields need "solid mechanics" physical fields, so these two physical fields need to be loaded in the simulation.

The cladding layer model was established in advance. In order to prevent the uncladding part from affecting the part under cladding, the "activated" node of COMSOL was used to simulate the laser cladding process during the numerical simulation calculation. Some parts of the field are deactivated before starting, while the deactivated part is multiplied by a small scaling factor to reduce the performance parameters of the material from the overall calculation. When the laser light source moves along the cladding path, the disused part is gradually activated and participates in the whole operation.

3. Results and Discussion

3.1. Formulation of Numerical Simulation Parameters

Figure 5 shows the temperature variation curves of horizontal sample points at different scanning speed, in Figure 5, the temperature in the center of the laser beam is higher than that on both sides. Because the sample points 10 and 11 are distributed symmetrically on both sides of the laser center, the temperature difference is not great at the same laser power and scanning speed. The laser power is proportional to the temperature of the cladding process and the maximum temperature increases with the increase of the laser power. The scanning speed is inversely proportional to the temperature of the cladding process, and the maximum temperature decreases with the increase of the scanning speed.

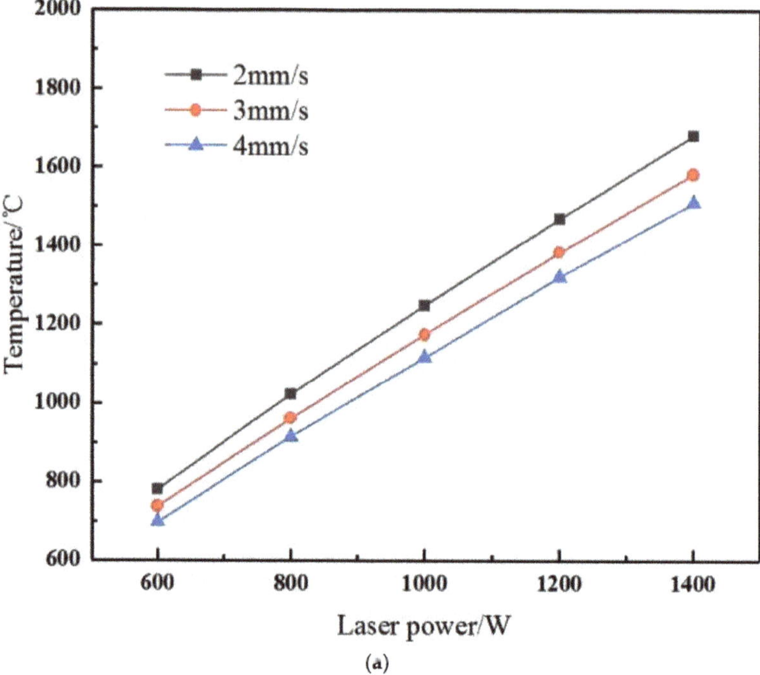

(a)

Figure 5. *Cont.*

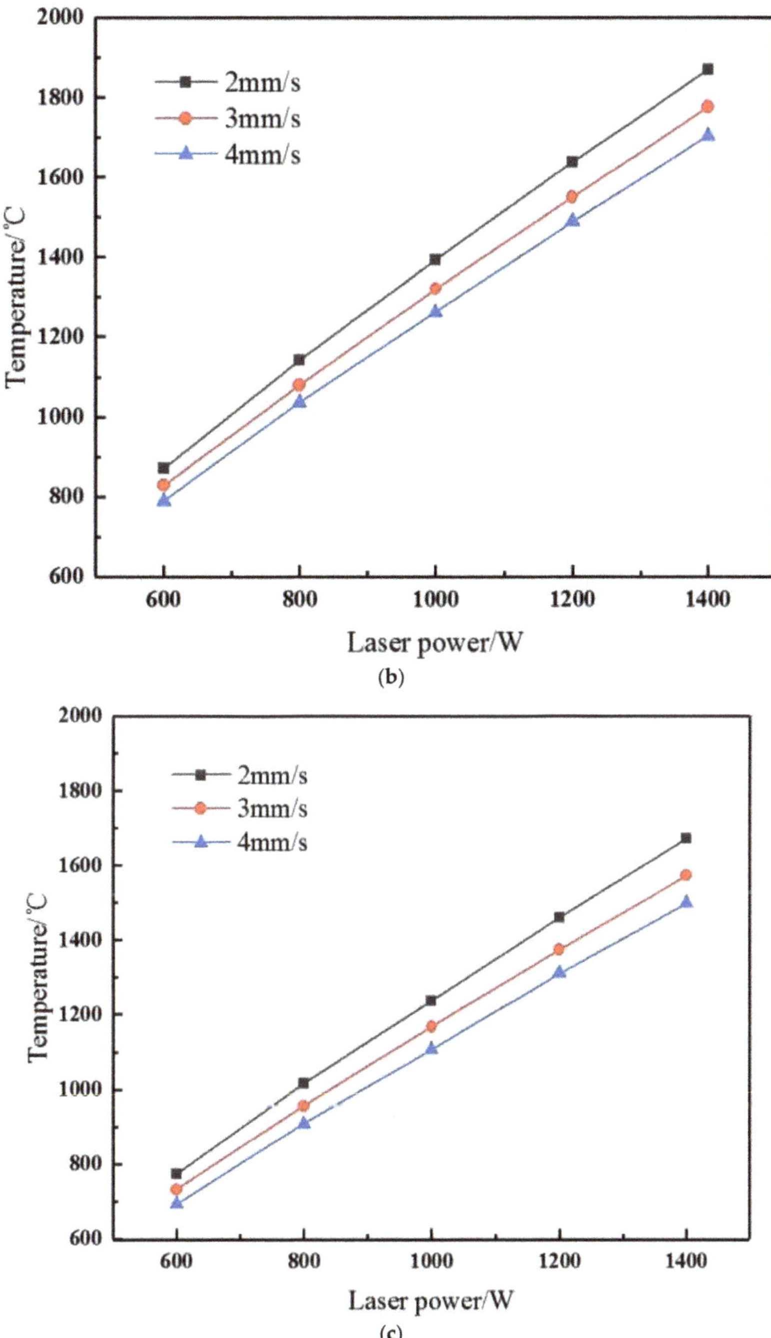

Figure 5. Maximum temperature variation with power of horizontal sample points at different scanning speed. (**a**) Sample point 10, (**b**) Sample point 6, (**c**) Sample point 11.

The three sample points are located on the matrix. If the maximum temperature of the three points reaches 1300 °C, it means that the cladding layer can be completely combined with the matrix. When the laser power is 600 and 800 W, the maximum temperature of the three places is less than 1300 °C, which does not meet the requirements; when the laser power is 1000 W, the maximum temperature can reach 1300 °C only when the scanning speed at B is 2 and 3 mm/s. When the laser power is 1200 W and 1400 W, the maximum temperature can reach the matrix melting temperature of 1300 °C. At this time, only the six schemes corresponding to the laser power of 1200 W and 1400 W meet the requirements.

3.2. Stress Field Analysis

Thermal stress is the main reason for cracks in the process of laser cladding, both laser power and scanning speed will affect the thermal stress value of the cladding layer. In order to further determine the numerical simulation parameter scheme, the thermal stress cloud images of the same position under six schemes are obtained through simulation (shown in Figure 6). The maximum thermal stress of the six schemes is listed in Table 3, it can be clearly observed that when the laser power is 1200 W and the scanning speed is 2 mm/s, the corresponding thermal stress value is the minimum. Therefore, this scheme is proposed as a numerical simulation scheme to study the thermal stress cycle.

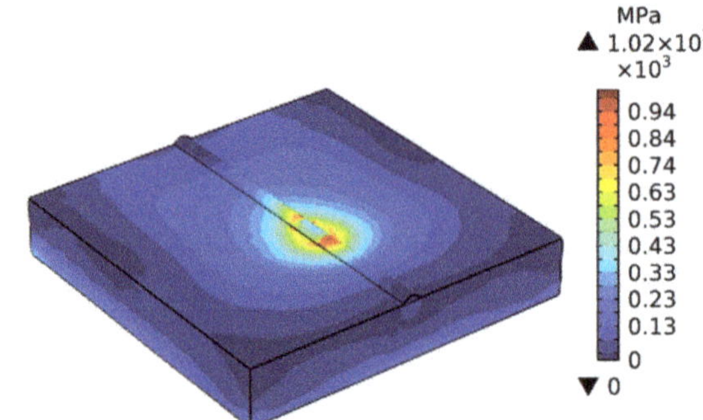

(**a**) Laser power 1200 W, scanning speed 2 mm/s, t = 14.5 s.

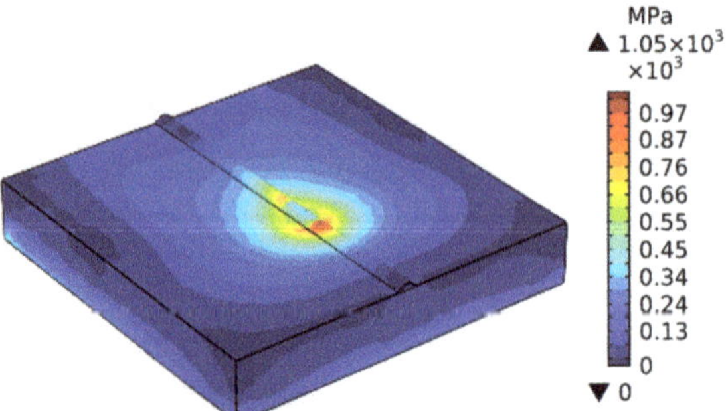

(**b**) Laser power 1200 W, scanning speed 3 mm/s, t = 20 s.

Figure 6. *Cont.*

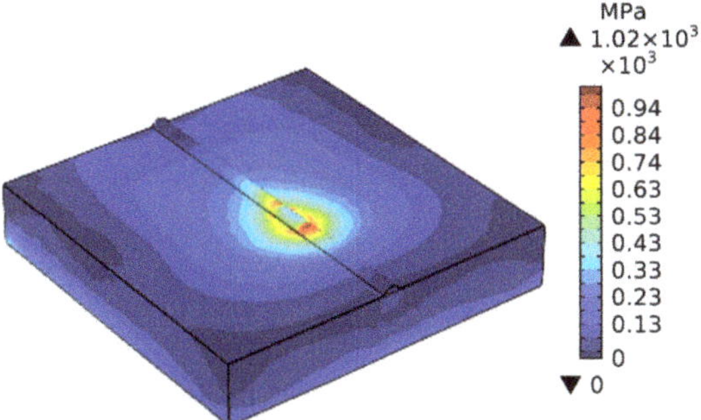

(**c**) Laser power 1200 W, scanning speed 4 mm/s, t = 18 s.

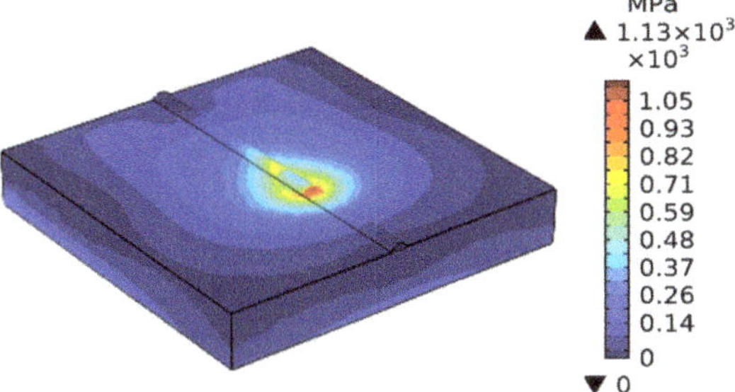

(**d**) Laser power 1400 W, scanning speed 2 mm/s, t = 14.5 s.

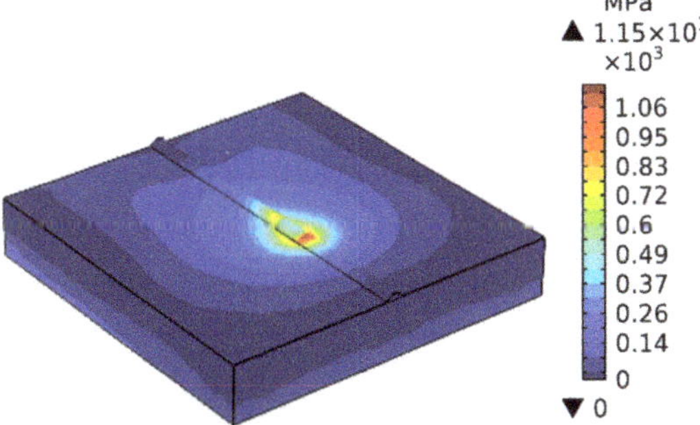

(**e**) Laser power 1400 W, scanning speed 3 mm/s, t = 20 s.

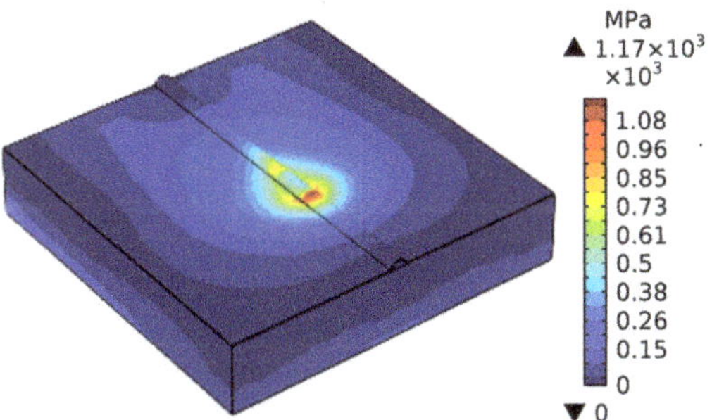

(**f**) Laser power 1400 W, scanning speed 4 mm/s, t = 14.5 s.

Figure 6. Von Mises thermal stress distribution at the same location under six schemes.

Table 3. Maximum thermal stress of von Mises.

Laser Power/W		1200			1400	
Scanning speed/(mm·s^{-1})	2	3	4	2	3	4
Maximum thermal stress of von Mises/MPa	1020	1050	1070	1130	1150	1170

Taking the optimal simulation scheme, the thermal stress and thermal cycle curves are drawn and analyzed as shown in Figure 7. In the figure, the sample points are heated rapidly by laser at 16 s and reach the maximum temperature at 17.5 s. In the vertical direction, the maximum temperature of all points decreases with the increase of the cladding layer depth. The melting point of powder is 1450 °C, while the highest temperature of sample points 1–5 is higher than the melting point of powder, so the powder can be completely melted. The melting point of the substrate is 1300 °C, while the maximum temperature of sample points 6 and 7 is higher than that of the substrate. The maximum temperature of sample point 8 is 1180 °C. Therefore, the junction between the cladding layer and the substrate is located between sample points 7 and 8, sample points 8~9 are the heat-affected zone of the cladding layer. The spacing between sample points is 0.2 mm, the depth of the molten pool is approximately 0.2~0.4 mm.

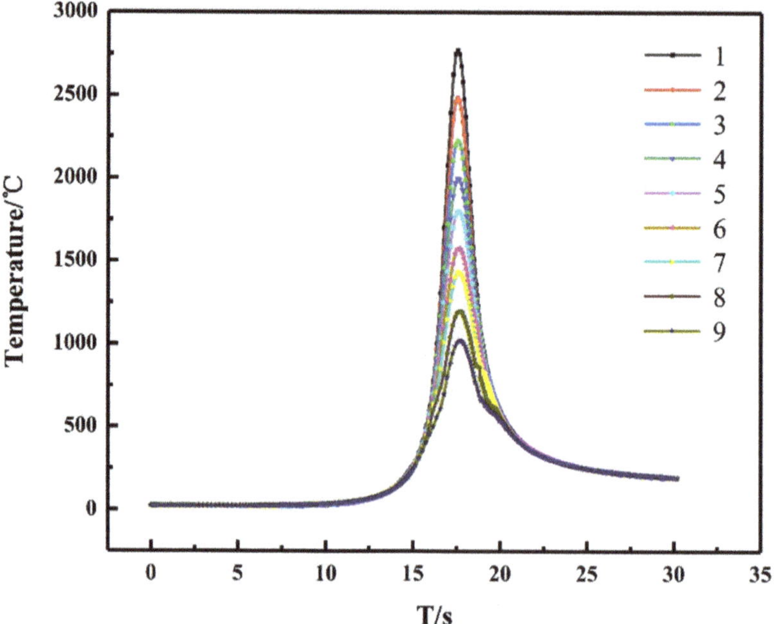

Figure 7. Thermal cycle curve.

Figure 8 shows the von mises thermal stress cycle, thermal stress cycle curves of most sample points have two peaks. When the laser beam is close to the sample point, the thermal expansion of the material around the molten pool exerts pressure on the sample point, so that the first peak of thermal stress occurs at the sample point, at which point the thermal stress of most sample points reaches its maximum value. When the laser beam is located directly above the sample point, the material at the sample point is melted, and the stress at the sample point rapidly decreases to the bottom of the valley. When the laser beam is far away from the sample point, the molten pool starts to solidify as the temperature drops and the stress increases gradually, then the second thermal stress peak appears at the sample point. As the temperature gradually decreases to room temperature, the thermal stress gradually decreases and finally tends to a stable value, namely the residual stress of the sample point.

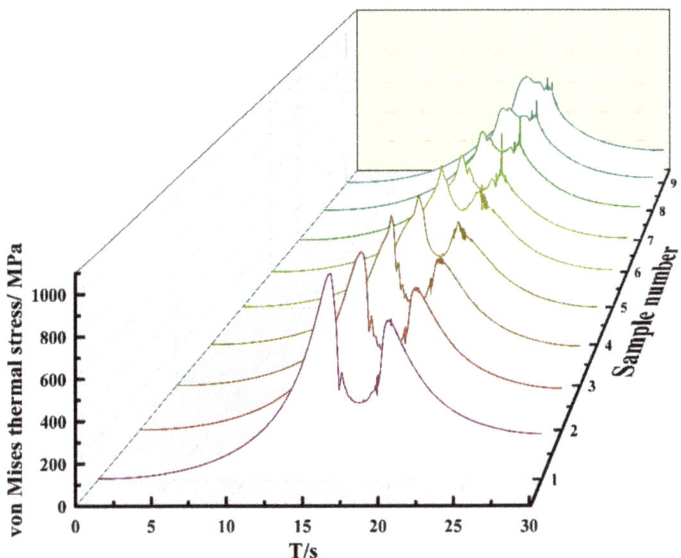

Figure 8. Von Mises thermal stress cycle.

In the vertical direction, with the increase of the depth cladding layer, the thermal stress at the two peak points gradually decreases, the thermal stress at the bottom of the cladding layer gradually increases, but the residual stress at each sample point tends to be the same value. The maximum thermal stress in the cladding process is 996.67 MPa, and the residual stress tends to be 210 MPa. The von Mises thermal stress curve does not have two distinct peaks when the sample point is outside the molten pool, because the material at the sample point did not melt. The shape of molten pool can be judged according to the thermal stress cycle curve. Sample points 8 and 9 in Figure 8 have no two obvious peak points. The lower side of the junction between the cladding layer and the matrix is located between sample points 7 and 8, which is consistent with the judgment of the thermal cycle curve.

Based on the von Mises thermal stress cycle, it was found that there were unstable alternating thermal stresses at each sample point. The characteristics of unstable alternating thermal stress are the unstable alternating thermal stress starts and ends at the same time, starting from 18.5 s and ending in 20 s, the unstable alternating thermal stress at sample points 1–4 will occur twice intensively, it is composed of multiple unstable alternating thermal stresses, there is a stable increase of thermal stress between the two times. With the increase of the cladding layer depth, the two unstable alternating thermal stresses gradually approach and connect together at sample point 5. As the cladding layer depth increases, the variation amplitude of alternating thermal stress increases first and then decreases, the maximum stress amplitude is 45.5 mpa.

3.3. Residual Stress Analysis

The analysis results of temperature and stress field are applied to the cladding layer as loads, and a line constraint is imposed on both sides of the matrix surface, the stress distribution of the workpiece in all directions after cooling for 3.5 h was analyzed, the stress is also the residual stress of the workpiece.

In order to study the residual stress distribution at different positions of the formed part, the middle position of the cladding layer was selected and cut along the plane. The junction between the cladding layer and the matrix along the axis is selected as the path1, and the path from the apex of the cladding layer to the vertical direction of the matrix along the axis is selected as the path2 to study the distribution of residual stress at different positions, as is shown in Figure 9.

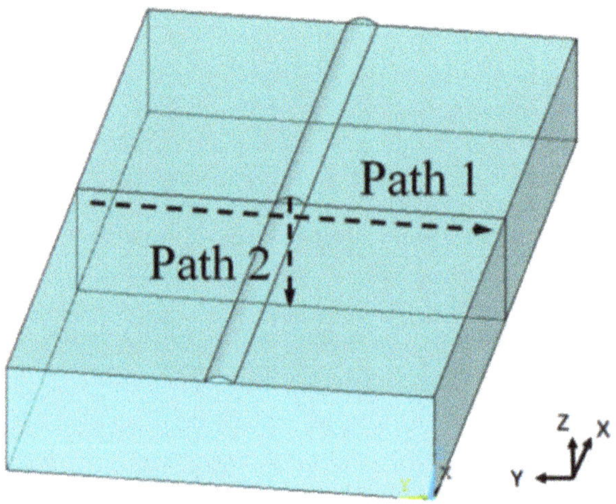

Figure 9. Path location diagram.

Figure 10 shows the residual stress distribution of the two paths. Figure 10a shows that in the horizontal direction of path 1, the residual stress is symmetrically distributed with a single track as the central axis. In X direction, the residual stress increases sharply when it is close to the boundary area of the cladding layer (0.7 mm $\leq y \leq$ 1.4 mm), and the maximum axial tensile stress is about 414 MPa. Because heat dissipation is relatively faster at the boundary between the cladding layer and the substrate, and the temperature gradient increases, thereby increasing the stress value. In the cladding layer, the closer to the center of the light spot, the smaller the temperature gradient, and the lower the stress value relative to the boundary of the cladding layer. Compressive stress is present in the part far away from the cladding layer, which is due to the high energy input, large temperature gradient, fast cooling rate, and difference in material properties during processing. After cooling, the substrate will hinder the shrinkage of the nickel-based coating, causing the cladding layer to bear tensile stress. Because of the limitation of the single-pass cladding width, the deformation resistance in the direction is small, and the stress value is also relatively small. However, the maximum residual stress in the y direction is concentrated on both sides of the cladding layer close to the substrate, cracks in the cladding layer often appear in this area. The tensile stress in Z direction is very small and can be ignored.

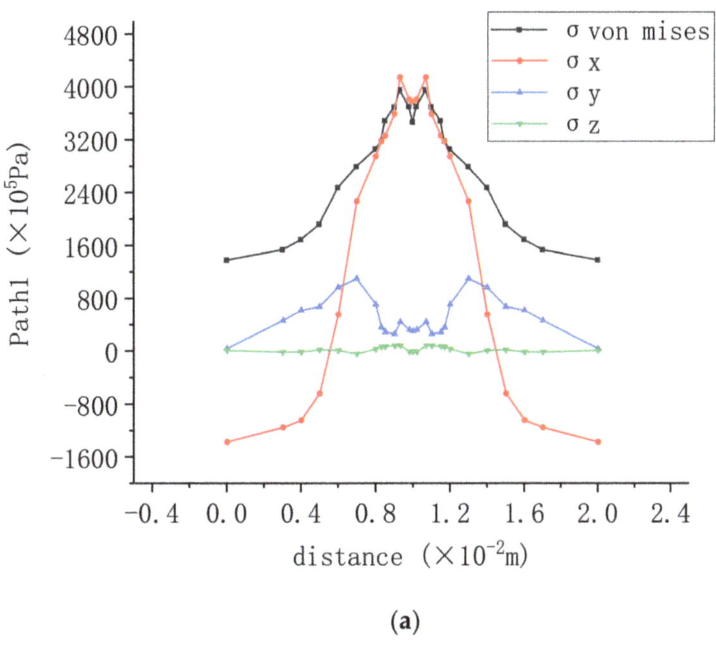

(a)

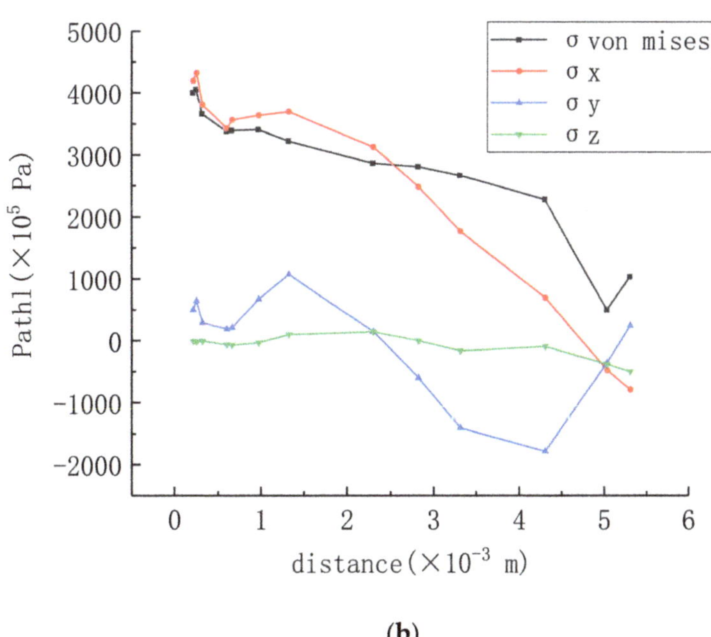

(b)

Figure 10. Residual stress distribution in two paths. (**a**) Residual stress distribution in path 1. (**b**) Residual stress distribution in path 2.

Figure 10b shows that there is a high tensile stress at the interface between the deposited layer and the substrate plane along the vertical direction of Path 2 due to the difference in thermophysical properties between substrate and nickel-based coating, and the maximum tensile stress at z = 3 mm is 432 Mpa. In x direction, as the depth continues to increase, the energy delivered by the laser becomes smaller and smaller, and the temperature gradient decreases, resulting in a gradual decrease in tensile stress. In y direction, as the depth of the cladding layer increases, the tensile stress gradually decreases and transforms into compressive stress in the middle of the substrate. But the compressive stress is transformed into tensile stress again at the bottom of the substrate z = 6 mm, because the local temperature is relatively high during the processing, and the temperature gradient of the entire cladding layer and its edges is large relative to the substrate, the shrinkage rate of the substrate is smaller than that of the entire single track. When the cladding layer is cooled and solidified, the substrates on both sides and the bottom of the cladding layer hinder the shrinkage of the cladding layer, so the single-pass cladding layer bears the tensile stress from the surrounding matrix. Because the cladding layer runs through the entire surface of the substrate along the axis, the overall deformation in the axial direction is an arc with a low center and high sides. Because the substrate is thicker, when the deposited layer is cooled and solidified, the compressive stress in the middle of the substrate and the tensile stress of the cladding layer balance each other according to the force balance principle, while the substrates on both sides move closer to the middle of the cladding layer, the bottom of the substrate bears the tensile stress. The residual stress value in z direction is relatively small, showing a small tensile stress in the upper position of the cladding layer and the substrate, it gradually shows a slight compressive stress as the depth increases.

3.4. Experimental Verification

In order to verify the accuracy of the simulation model, the simulation scheme proposed above was taken for laser cladding experiment. In Figure 11, the temperatures of sample point 6 and 7 are respectively 1580 °C and 1460 °C, which are higher than the melting point of matrix 1300 °C, and the temperature of sample point 8 is 1180 °C. Therefore, according to the simulation calculation, it can be judged that the junction of cladding layer and matrix is located between sample points 7 and 8. After measurement, the depth, width and height of the cladding layer are 0.28 mm, 3.0 mm, and 1.0 mm respectively. The height and width of the cladding layer set by the finite element model are 1.0 mm and 3 mm, the depth of the molten pool obtained by the simulation calculation is 0.2~0.4 mm. It can be seen that the actual size of the cladding layer is basically consistent with the simulation results. The cross section morphology of the cladding layer is basically the same by comparing the cross section screenshot with the simulation results, which verifies that the simulation model is correct.

Figure 12 shows that obvious cracks appear on both sides of the cladding layer close to the substrate, which also has the maximum residual stress according to the simulation results. But the cracks are very small, because the load parameters of the residual stress simulation are optimized. The cracks can only be improved, but cannot be absolutely eliminated.

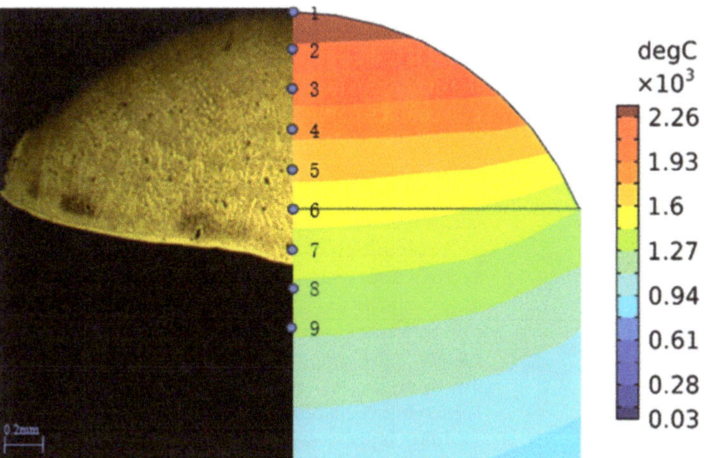

Figure 11. Experimental and simulation results.

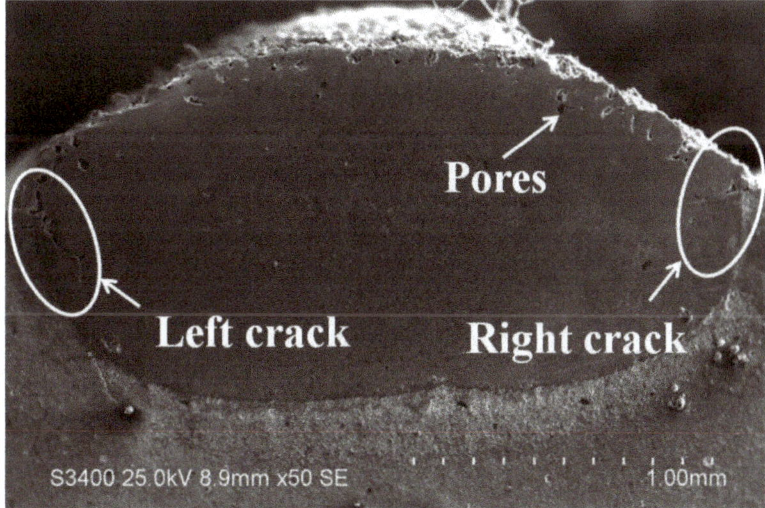

Figure 12. Crack location of cladding layer cross section.

4. Conclusions

1. The numerical simulation of Ni-base alloy powder laser cladding on H13 steel was carried out. The optimal process parameters are as follows: laser power is 1200 W, scanning speed is 2 mm/s. Under these parameters, the maximum temperature of laser cladding is about 2748.1 °C, the maximum heating rate is 1632.1 °C/s, the maximum cooling rate is 699.5 °C/s, the depth of molten pool is 0.28 mm. Under these parameters, the cross section of the cladding layer obtained by laser cladding experiment is basically consistent with the simulation results, which proves the correctness of the simulation model.
2. The temperature field and thermal stress field of the cladding layer were analyzed. The laser center temperature of the cladding layer at the same horizontal plane is greater than that on both sides. The temperature difference between the two sides of

the laser center is not so great. The laser power is proportional to the temperature of cladding, the maximum temperature increases with the laser power. The scanning speed is inversely proportional to the maximum temperature during cladding, and the maximum temperature decreases with the increase of scanning speed. The laser power and scanning speed are proportional to the thermal stress at the sample point and the thermal stress increases with the increase of laser power and scanning speed. For the von Mises thermal stress cycle curve, the thermal stress curve of most samples has two peaks. This is mainly due to the solid-liquid phase transition process in laser cladding, but when the sample point is outside the molten pool, because the material at the sample point is not melted, then the corresponding peak value of the von Mises thermal stress curve is not obvious. Based on this feature, the depth of the molten pool can be predicted.

3. The residual stress of the cladding layer was simulated according to the analysis results of the temperature and stress field. The cladding layer mainly bears residual tensile stress, because its cooling shrinkage is inhibited by the matrix. On the cross section of the cladding layer, the maximum tensile stress appears on both sides of the cladding layer close to the substrate, where cracks always appear.
4. The cladding experiment investigates that the simulation results of von Mises thermal stress cycle and residual stress are correct.

Author Contributions: Conceptualization, F.Y.; methodology, F.Y.; software, F.Y.; validation, L.F.; formal analysis, F.Y.; investigation, F.Y.; data curation, F.Y.; writing—original draft preparation, F.Y.; writing—review and editing, F.Y.; visualization, F.Y., L.F. All authors have read and agreed to the published version of the manuscript.

Funding: This research received no external funding.

Institutional Review Board Statement: Not applicable.

Informed Consent Statement: Not applicable.

Conflicts of Interest: The authors declare no conflict of interest.

References

1. Cao, J.; Lu, H.F.; Lu, J.Z. Effects of Tungsten Carbide Particles on Microstructure and Wear Resistance of Hot-Working Die Prepared via Laser Cladding. *Chin. J. Lasers* **2019**, *46*, 702001.
2. Cui, C.; Wu, M.P.; Xia, S.H. Effect of Heat Treatment on Properties of Laser Cladding Cobalt-Based Coating on 42CrMo Steel Surface. *Chin. J. Lasers* **2020**, *47*, 602011.
3. Li, G.S.; Li, J.H.; Feng, W.L. Effects of specific powder and specific energy on the characteristics of NiWC25 by laser cladding. *Surf. Technol.* **2019**, *48*, 253–258.
4. Li, J.H.; Li, G.S.; Zhang, D.Q. Study on microhardness of Laser cladding NJ-4 powder. *Surf. Technol.* **2018**, *47*, 77–83.
5. Chen, H.; Lu, Y.; Sun, Y.; Wei, Y.; Wang, X.; Liu, D. Coarse TiC particles reinforced H13 steel matrix composites produced by laser cladding. *Surf. Coat. Technol.* **2020**, *395*, 125867. [CrossRef]
6. Zhao, S.J.; Qi, W.J.; Hang, Y.F. Numerical simulation study on thermal cycle characteristics of temperature field TC4 surface laser cladding Ni60 base coating. *Surf. Technol.* **2020**, *49*, 301–308.
7. Wang, X.; Yu, Y.G.; Hang, E.Z. Research status of laser cladding temperature field simulation. *Therm. Spray Technol.* **2019**, *11*, 10–15.
8. Muvvala, G.; Mullick, S.; Nath, A.K. Development of process maps based on molten pool thermal history during laser cladding of Inconel 718/TiC metal matrix composite coatings. *Surf. Coat. Technol.* **2020**, *399*, 126100. [CrossRef]
9. Kong, F.; Kovacevic, R. Modeling of heat transfer and fluid flow in the laser multilayered cladding process. *Metall. Mater. Trans. B* **2010**, *41*, 1310–1320. [CrossRef]
10. Liu, H.; Yu, G.; He, X.L. Three-dimensional numerical simulation of transient temperature field and coating geometry in powder feeding laser cladding. *Chin. J. Lasers* **2013**, *40*, 1203007. [CrossRef]
11. Xu, W.F.; Ma, J.; Luo, Y.X. Microstructure and high-temperature mechanical properties of laser beam welded TC4/TA15 dissimilar titanium alloy joints. *Trans. Nonferrous Met. Soc. China* **2020**, *30*, 160–170. [CrossRef]
12. Song, B.; Yu, T.; Jiang, X.; Xi, W.; Lin, X. Effect of laser power on molten pool evolution and convection. *Numer. Heat Transf. Part A Appl.* **2020**, *78*, 1777795. [CrossRef]
13. He, S.Y.; Liu, X.D.; Zhao, S.Z. Microstructure and Wear Resistance of Carbon Fibers Reinforced 316L Stainless Steel Prepared Using Laser Cladding. *Chin. J. Lasers* **2020**, *47*, 0502010.

14. Song, P.F.; Jiang, F.L.; Wang, Y.L. Advances in the preparation of high entropy alloy coatings by laser cladding. *Surf. Technol.* **2020**, *1*, 18.
15. Deirmina, F.; Peghini, N. Heat treatment and properties of a hot work tool steel fabricated by additive manufacturing. *Mater. Sci. Eng. A* **2019**, *753*, 109–121. [CrossRef]
16. Zhu, J.; Zhang, Z. Improving strength and ductility of H13 die steel by pre-tempering treatment and its mechanism. *Mater. Sci. Eng. A* **2019**, *752*, 109–114. [CrossRef]
17. Wang, Y.; Song, K. Microstructure evolution mechanism near the fracture lip of $4Cr_5MoSiV_1$ steel during deforming at 580 °C. *J. Mater. Res. Technol.* **2019**, *8*, 6390–6395. [CrossRef]
18. Krylova, S.E.; Romashkov, E.V. Special aspects of thermal treatment of steel for hot forming dies production. *Mater. Today Proc.* **2019**, *19*, 2540–2544. [CrossRef]
19. Kar, A.; Mazumder, J. One-dimensional diffusion model for extended solid solution in laser cladding. *J. Appl. Phys.* **1987**, *61*, 2645–2655. [CrossRef]
20. Hoadley AF, A.; Rappaz, M. A thermal model of laser cladding by powder injection. *Metall. Trans. B* **1992**, *23*, 631–642. [CrossRef]
21. Han, L.; Phatak, K.M.; Liou, F.W. Modeling of laser cladding with powder injection. *Metall. Mater. Trans. B* **2004**, *35*, 1139–1150. [CrossRef]
22. Cho, C.D.; Zhao, G.P.; Kwak, S.Y. Computational mechanics of laser cladding process. *J. Mater. Process. Technol.* **2004**, *153–154*, 494–500. [CrossRef]
23. Jendrzejewski, R.; Sliwinski, G.; Krawczuk, M. Temperature and stress fields induced during laser cladding. *Comput. Struct.* **2004**, *82*, 653–658. [CrossRef]
24. Toyserkani, E.; Khajepour, A.; Corbin, S. 3-D finite element modeling of laser cladding by powder injection: Effects of laser pulse shaping on the process. *Opt. Lasers Eng.* **2004**, *41*, 849–867. [CrossRef]
25. He, X.; Mazumder, J. Transport phenomena during direct metal deposition. *J. Appl. Phys.* **2007**, *101*, 53113. [CrossRef]
26. He, X.; Yu, G.; Mazumder, J. Temperature and composition profile during double-track laser cladding of H13 tool steel. *J. Phys. D Appl. Phys.* **2010**, *43*, 015502. [CrossRef]
27. Farahmand, P.; Kovacevic, R. An experimental-numerical investigation of heat distribution and stress field in single-and multi-track laser cladding by a high-power direct diode laser. *Opt. Laser Technol.* **2014**, *63*, 154–168. [CrossRef]

Article

Improvement of Wear, Pitting Corrosion Resistance and Repassivation Ability of Mg-Based Alloys Using High Pressure Cold Sprayed (HPCS) Commercially Pure-Titanium Coatings

Mohammadreza Daroonparvar [1,*], Ashish K. Kasar [2], Mohammad Umar Farooq Khan [3], Pradeep L. Menezes [2,*], Charles M. Kay [4], Manoranjan Misra [1] and Rajeev K. Gupta [3]

1 Department of Chemical and Materials Engineering, University of Nevada, Reno, NV 89501, USA; misra49@gmail.com
2 Department of Mechanical Engineering, University of Nevada, Reno, NV 89501, USA; akasar@nevada.unr.edu
3 Department of Materials Science and Engineering, College of Engineering, North Carolina State University, 911 Partners Way, Raleigh, NC 27695, USA; mk189@zips.uakron.edu (M.U.F.K.); rkgupta2@ncsu.edu (R.K.G.)
4 ASB Industries Inc., Barberton, OH 44203, USA; cmkay@asbindustries.com
* Correspondence: mr.daroonparvar@yahoo.com (M.D.); pmenezes@unr.edu (P.L.M.)

Abstract: In this study, a compact cold sprayed (CS) Ti coating was deposited on Mg alloy using a high pressure cold spray (HPCS) system. The wear and corrosion behavior of the CS Ti coating was compared with that of CS Al coating and bare Mg alloy. The Ti coating yielded lower wear rate compared to Al coating and Mg alloy. Electrochemical impedance spectroscopy (EIS) and cyclic potentiodynamic polarization (CPP) tests revealed that CS Ti coating can substantially reduce corrosion rate of AZ31B in chloride containing solutions compared to CS Al coating. Interestingly, Ti-coated Mg alloy demonstrated negative hysteresis loop, depicting repassivation of pits, in contrast to AZ31B and Al-coated AZ31B with positive hysteresis loops where corrosion potential (E_{corr}) > repassivation potential (E_{rp}); indicating irreversible growth of pits. AZ31B and Al-coated AZ31B were most susceptible to pitting corrosion, while Ti-coated Mg alloy indicated noticeable resistance to pitting in 3.5 wt % NaCl solution. In comparison to Al coating, Ti coating considerably separated the AZ31BMg alloy surface from the corrosive electrolyte during long term immersion test for 11 days.

Keywords: Ti coating; cyclic potentiodynamic polarization (CPP) test; hysteresis loop; wear

Citation: Daroonparvar, M.; Kasar, A.K.; Farooq Khan, M.U.; L. Menezes, P.; Kay, C.M.; Misra, M.; Gupta, R.K. Improvement of Wear, Pitting Corrosion Resistance and Repassivation Ability of Mg-Based Alloys Using High Pressure Cold Sprayed (HPCS) Commercially Pure-Titanium Coatings. *Coatings* **2021**, *11*, 57. https://doi.org/10.3390/coatings11010057

Received: 16 December 2020
Accepted: 28 December 2020
Published: 6 January 2021

Publisher's Note: MDPI stays neutral with regard to jurisdictional claims in published maps and institutional affiliations.

1. Introduction

Magnesium (Mg) and its alloys (with a low density of about 1.8 g/cm^3) have become a hot topic of research because of the potential engineering applications. Moreover, magnesium alloys show outstanding potential in automotive, aerospace, and electronic industries because of their high strength-to-weight ratio, high stiffness, outstanding electromagnetic shielding ability, and remarkable damping performance, etc. In recent years, Mg alloys have been receiving ascending attention as biodegradable implant materials, as well. Regrettably, the inferior wear and corrosion performances of Mg alloys severely limit their extensive applications. The most commonly employed method for improving the surface properties of a substrate is surface treatment. In this regard, various conversion coatings [1], anodization process, plasma electrolytic oxidation (PEO), physical vapor deposition (PVD), electro-less, before processes: annealing processes, electroplating and ions implantation methods and thermal spray processes have been employed to modify the surface of Mg alloys for improving their wear and corrosion resistances [2–10].

An approximate new coating technology that deserves particular attention is cold spray process (as an environmentally friendly method) which doesn't involve toxic fumes or other harmful emissions [11,12]. Compared to the high velocity oxy-fuel (HVOF) thermal spray process which uses a combination of thermal and kinetic energies, cold spray utilizes only kinetic (dynamic) energy to deposit the powder particles [12–15]. Likewise, the microstructural degeneration of heat-susceptible substrates such as Mg alloys which is frequently seen in the substitute thermal spray methods could be prevented by means of cold spray process [4,16–18]. In contrast to thermal spray technologies such as electric arc wire spray, plasma spray, flame spray and HVOF spray processes which partially and/or fully melt particles during the spray process; CS can avert the thermal effects including oxidation, porosity, grain growth and phase transformation during spray process [12,13,19,20].

It was reported that corrosion resistance of Mg alloys can be improved with the cold sprayed coatings (in comparison with counterpart coatings made by other techniques e.g., anodizing, E-plating, conversion coating and etc.), in 3.5 wt % NaCl solution [21]. Aluminum (with good corrosion resistance, low density, and having low standard electrode potential difference with Mg alloys) is used (as protective coating) to reduce the corrosion rate of Mg substrates [2,22–24]. Current cold sprayed (N_2 as propellant gas) Al-based coatings (as single layer) on Mg alloys lack acceptable hardness, wear resistance and are highly susceptible to localized corrosions in severe corrosive atmospheres [22,25,26]. These coatings also showed low repassivation ability [27]. In fact, the passive film has a weak propensity to repair itself (or passivate) in corrosive environment.

Compared to Al and its alloys, Ti and its alloys can be extensively used in severe corrosive environments such as offshore (salt water), aerospace, automotive, etc. This was attributed to the good mechanical properties and excellent corrosion resistance (due to the formation of a firm protective oxide film on the metal surface) [28–30]. Low standard reduction potential mismatch between coating and substrate makes Ti coating (from group 4B) as a subsequent candidate for the corrosion protection of Mg and its alloys [31,32]. In this regard, warm sprayed (WS) Ti coatings couldn't noticeably enhance the corrosion potential and lower the corrosion current density of AZ91E Mg alloy [31]. These coatings disclosed poor corrosion resistance and finally led to the fast degeneration of Mg alloy. The poor performance of WS Ti coatings was due to the presence of through-thickness porosities which simply conducted the chloride containing solution towards the substrate surface. The untimely tear of titanium coatings (after only 24 h of immersion) in 3.5 wt % NaCl electrolyte was eventually observed [31]. This was mainly related to the corrosion products formation and accumulation at the interface between WS Ti coating and Mg substrate [31]. The MS (magnetron sputtered) Ti-coated AZ91D Mg alloy showed even much inferior performance than the uncoated AZ91D Mg alloy after 1 day in NaCl solution [32]. Most part of MS Ti coatings came off the Mg alloy substrate which had undergone the severe corrosion [32].

In this research, we developed a fairly compact cold sprayed titanium coating on Mg alloy using HPCS system. It is anticipated that high pressure cold sprayed commercially pure-Ti coating could alleviate the problems associated with current cold sprayed Al coatings on Mg alloys and exceptionally increase the repassivation ability of Mg alloys. Moreover, immersion test for 11 days was performed to further elucidate the effectiveness and corrosion protection performance of HPCS titanium coating on magnesium alloys in corrosive environment.

2. Experimental Methods

2.1. Feedstock Powders and Substrate

In this research, commercially pure (CP) Al, and CP-Ti grade 1 powders (as feedstock powders) were employed for coating production. Commercially available AZ31B Mg alloy plate (381 mm × 455 mm × 9.5 mm) was procured from Magnesium EleKtronNorth America. The substrates were then cut from this plate. Table 1 shows the chemical composition of AZ31B Mg alloy. The substrate surface was grit blasted and then cleaned with alcohol and acetone using a hand spray bottle and N_2 blow off right before cold spray process.

Table 1. Chemical composition of AZ31B (wt %).

Alloy Name	Mg	Al	Zn	Si	Mn	Fe	Cu	Ca	Ni	C
AZ31B Mg alloy	Balance	2.8	1.1	<0.08	0.7	<0.01	<0.01	<0.03	<0.001	-

2.2. CS Deposition

In this research, high pressure cold spray system (Impact Innovation 5/11 system, GmbH,) was used to produce metallic layers on AZ31BMg substrates. The temperature of substrate and coatings was maintained less than 65 °C during cold spray process. Likewise, Table 2 displays the cold spray parameters. The following coatings were sprayed on the AZ31B Mg alloy:

1. CS Al coating on AZ31B Mg.
2. CS Ti coating on AZ31B Mg.

Table 2. CS process parameters.

Propellant Gas	Sprayed Material	Gas Temperature (C°)	Gas Pressure (MPa)	Spray Angle (°)	Stand-Off Distance (mm)	Step Size (mm)	Powder Feed Rate (RPM)	Powder Carrier Gas Flow Rate (m^3/hr)	Type of Nozzle
N_2	CP-Al	350–600	3.0–5.0	90°	25.4	0.508	2.0–3.5	3.0	SiC water cooled nozzle
N_2	CP-Ti	700–1050	3.0–5.0	90°	25.4	1.016	2.0–3.5	3.0	SiC water cooled nozzle

2.3. Characterization

Optical microscopy (IX70, Olympus) was used to analyze the polished cross-sectional microstructures of as-sprayed coatings on the AZ31B. For this purpose, coated AZ31B samples were cut, mounted and polished using with standard metallographic procedures on an Allied Metprep 3^{TM} grinder/polisher system. ImageJ software was utilized to analyze the porosity level of the as-sprayed coatings (using ASTM E2109-01) [33]. The surface morphology, polished cross-sectional microstructures and chemical composition (elemental analysis) of developed coatings before and after immersion test were studied using a LYRA-3 model XUM integrated variable pressure focused ion beam-field emission scanning electron microscope (FIB-FESEM; TESCAN), and a TM-3030 Scanning Electron Microscope (SEM; Hitachi,) equipped with Energy Dispersive X-ray spectroscopy (EDS) capability. Moreover, the structural phases of as-sprayed coatings, bare AZ31B Mg alloy substrate and feed stock powders were analyzed using an Ultima IV X-ray machine (Rigaku), after grinding up to 1200 grit size sandpaper (for only bare and coated Mg alloys). The X-ray tube emits Cu-Kα radiation with an excitation voltage of 40 kV and excitation current of 35 mA. The samples were scanned at a rate of 1 degree/min with a step width of 0.04 degree. The data were analyzed using X'Pert HighScore Plus software with ICDD database. Furthermore, 2θ (as diffraction angle) range of 30°–90° was employed to collect diffraction patterns of the different samples.

A Vickers hardness tester (Beuhler-Wilson Tukon 1202), was used to measure the micro-hardnesses of the substrate and the coatings, under the load of 0.245 N. It should be noted that the substrate hardness measurements were performed at the regions away from the interface between coating and the Mg alloy substrate. Additionally, ten (10) measurements were done on each sample and the average was reported as micro-hardness value.

Average surface roughness (R_a) of as-cold sprayed coatings were inspected during profilometry using an Alicona Infinite Focus, a 3D measurement system which has a non-contact, optical measurement principle based on focus-variation. Prior to profilometry, the surface was cleaned with DI water using an ultrasonic cleaner. The brightness and contrast were adjusted at a range of focus to make sure all the features are within focus during the scan. The lateral resolution was set at 50 nm. R_a was measured using line scans across the IFM scan. At least five readings were collected for R_a to minimize standard deviation.

Dry reciprocating sliding tests (according to ASTM G133-05 [34]) were performed using a Rtech-Tribometer at room temperature (~25 °C and 40–50% relative humidity).

Prior to sliding tests, all the coated surfaces were polished to achieve an average surface roughness (R_a) of 0.2 ± 0.05 μm. R_a is defined as the arithmetic mean of the absolute values of the vertical deviation from the mean line through the profile [35]. E52100 steel ball with 6.35 mm diameter was used as a counterpart. All the reciprocating sliding tests were carried out with a track length of 15 mm and 1 mm/s velocity under a normal load of 4 N for a total distance of 1000 mm. The 1000 mm of sliding distance was selected based on the stabilized wear depth during sliding. The wear depth was recorded to measure the wear volume. The specific wear rate was then measured using the following formula [36]:

$$\text{Wear rate } \left(\mu\text{m}^3/\text{Nm}\right) = \frac{\textit{Wear volume } \left(\mu\text{m}^3\right)}{\textit{Normal load } (\text{N}) \times \textit{Sliding distance } (\text{m})} \tag{1}$$

2.4. Sample Preparation for Corrosion Tests

The surface area of as-cold sprayed coatings which have a highly active surface [37] is increased. This is attributed to the rough surface of as-cold sprayed coatings. As a general practice, the rough and porous surface layer of as-cold sprayed coatings should be removed before corrosion tests [28]. Hence, the samples surface was ground up to 1200 US grit size sandpaper (SiC abrasive papers) and then cleaned with ethanol using an ultrasonic cleaner for 5 min before cyclic potentiodynamic polarization (CPP), electrochemical impedance spectroscopy (EIS) and long term immersion tests.

2.5. Cyclic Potentiodynamic Polarization (CPP) Tests in 3.5 wt % NaCl Solution

Cyclic potentiodynamic polarization tests were performed in a three-electrode setup using a flat cell and a Bio-logic potentiostat (per ASTM standard G61 [38]). The standard calomel electrode, platinum electrode, and sample under test were connected as a reference electrode, counter electrode, and working electrode, respectively. Before the CPP test, the open circuit potential (OCP) was tracked for 1 h to allow the system to achieve an equilibrium in the electrolyte. PDP tests were done in 3.5 wt % NaCl (pH 6.7) at a scan rate of 1 mV/s from 200 mV below OCP to a current limit of 10 mA/cm^2 or a potential limit of 2.5 V_{SCE} and reversed back to the same starting potential of 200 mV below OCP at room temperature. The pitting potential was determined by intersecting the line coming from extending the passive current density region and the linear increase in the current density region after passivation region.

2.6. Electrochemical Impedance Spectroscopy (EIS) in 3.5 wt % NaCl Solution

A three-electrode setup was used in a flat cell where standard calomel electrode, platinum electrode, and sample were connected as a reference electrode, counter electrode, and working electrode, respectively. The sample was monitored for 1 h observing the OCP in 3.5% NaCl (pH 6.7) exposing an area of 1 cm^2 at room temperature. Likewise, 100 kHz to 10 mHz (as frequency range) at OCP was selected for performing EIS test. For each EIS scan, ten measurements were recorded per decade, with an average of at least three points per measurement. Furthermore, sinusoidal AC perturbation with an amplitude of 10 mV (rms) was considered for all EIS tests. EC-lab 11.21 provided in the Bio-logic potentiostat was utilized to analyze CPP as well as the EIS data. Electrochemical corrosion tests were carried out three times to substantiate the repeatability of the obtained results. After the immersion test for 11 days, the samples were rinsed with DI water and subsequently dried in air.

3. Results and Discussion

3.1. Powders Morphology and Coatings Microstructure

CP (commercially pure)-Al (in the size range of 9–40 μm), and CP-Ti (in the size range of 10–45 μm) powders all possess spherical morphology as shown in Figure 1a,b respectively. Surface morphology of CS Ti and Al coatings (in as-sprayed condition) is shown in Figure 2b,d, respectively. R_a of coatings (in as-sprayed condition) was 2.816 ± 0.7 μm and

2.182 ± 0.7 μm for Al and Ti coatings, respectively. Lower R_a for Ti coating was due to intense plastic deformation of Ti particles during CS process. Figures 2a,c and 3 demonstrate the microstructure of polished cross section of the coatings on the AZ31B substrate.

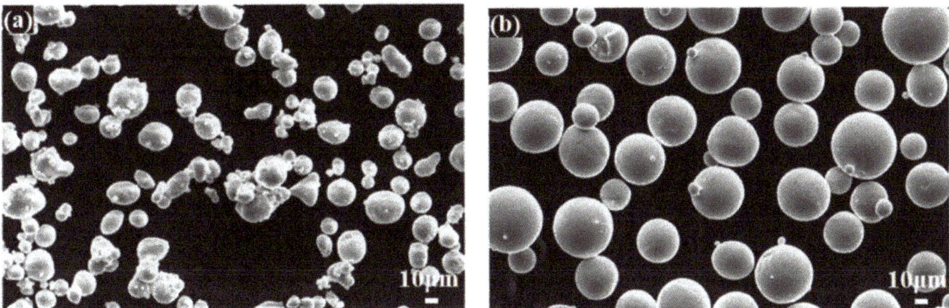

Figure 1. SEM images of feedstock powders (**a**) commercially pure-Al, and (**b**) commercially pure-Ti powders.

The local deformation of Ti powder particles was obvious (Figure 3c–e). Moreover, relatively dense microstructure (Figure 3a,b and Figure 2a) along with very limited micropores with porosity level of about 0.40 ± 0.20% was observed for titanium coating (in this research work). On the contrary, higher level of porosities were detected in atmospheric plasma sprayed (APS) Ti coating (with 10.2% porosity level) and CS Ti coating (with 2.7% porosity level) as well [31,39]. Different thermal spray methods have been employed to deposit Ti coatings with low level of porosities and high purity as well. However, thermal degradation of the deposited Ti powder particle occurs during HVOF spray process. This could be related to the temperature range of spray powder particles which is about 1227–2427 °C. In fact, the probability of a hard and brittle oxygen enriched layer formation (in the case of Ti) which is also known as "α-case" is expected and could cause the notable loss of plasticity, ductility, etc. [31]. Another method is LPPS (low pressure plasma spray) process that could restrict oxidation during spray process. This is attributed to the vacuum environment which makes this technique costly. Instead, in warm spray (as modification of HVOF spray system [31]) method, supersonic gas flow temperature is adjusted by injecting N_2 into the mixing chamber. Formation of a relatively dense coating with limited oxidation of powder particles is anticipated using this method. Nevertheless, higher level of porosities was reported in the deposited Ti coatings (with porosity level of about 3.8–5.5%) on Mg alloys by WS method under different N_2 flow rates [31,39]. Higher level of porosities in warm sprayed Ti coatings substantially declined their corrosion resistance in 3.5 wt % NaCl solution and caused quick degradation of Mg substrate during long term corrosion (after 24h of immersion) [31].

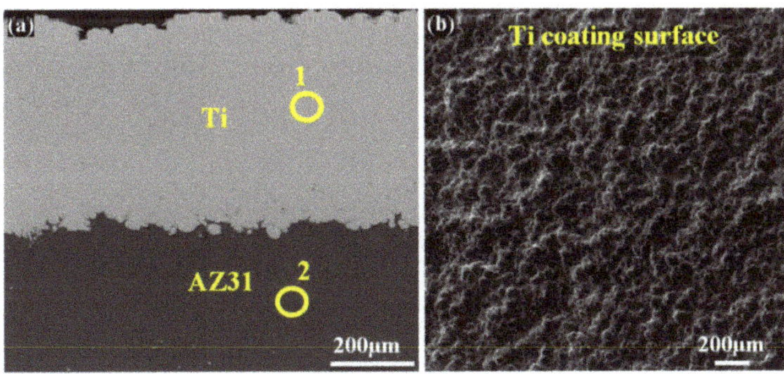

Figure 2. *Cont.*

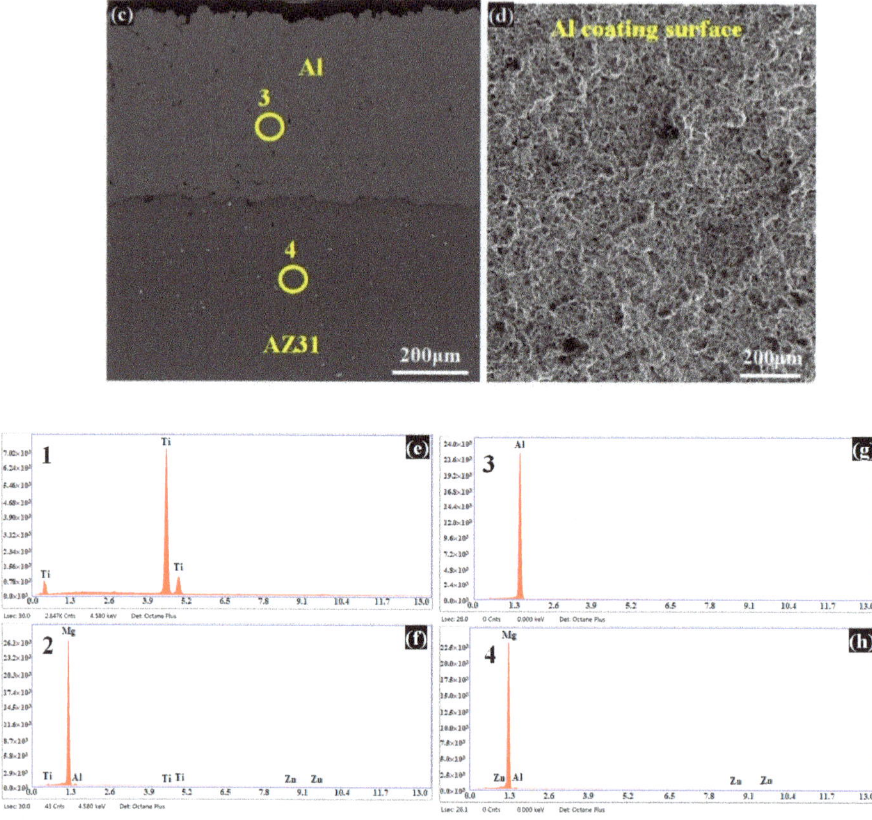

Figure 2. SEM images from (**a**) cross section of Ti-coated AZ31B, (**b**) Ti coating surface, (**c**) cross section of Al-coated AZ31B, (**d**) Al coating surface, EDS analysis of 1 (**e**), 2 (**f**), and EDS analysis of 3 (**g**), 4 (**h**).

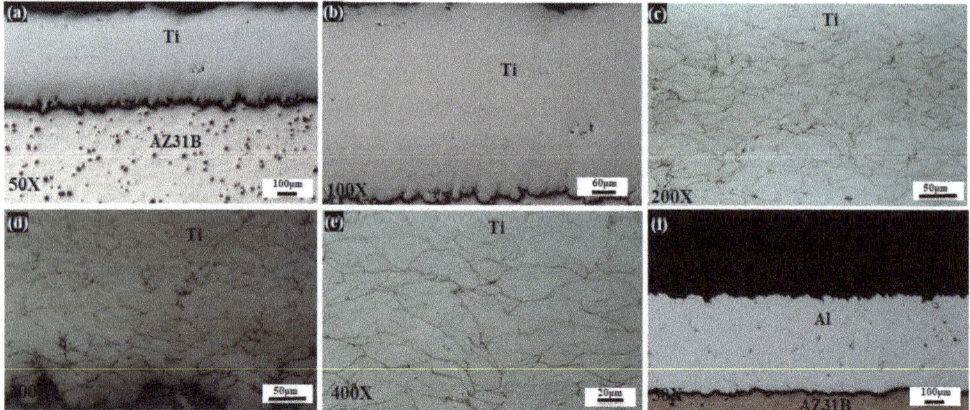

Figure 3. *Cont.*

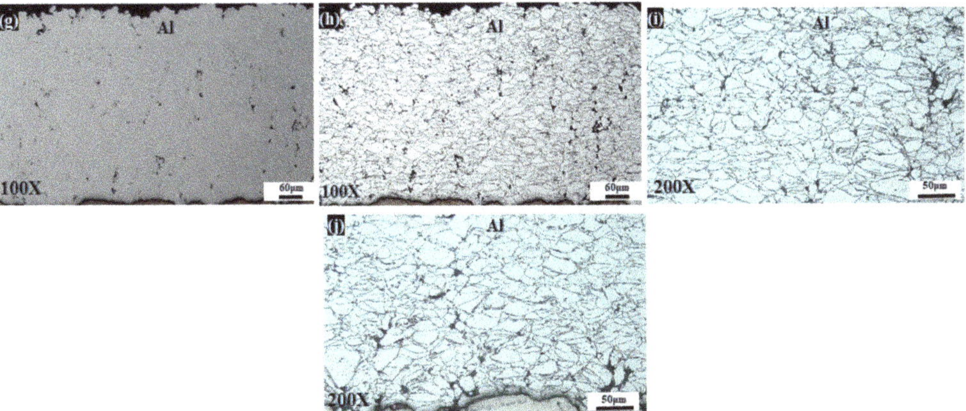

Figure 3. Photomicrographs from polished crossed section of (**a–e**) Ti-coated AZ31B at different magnifications, and (**f–j**) Al-coated AZ31B at different magnifications before and after etching.

In this study, CP-Al coating with porosity level of about 1.00 ± 0.20% depicted higher porosity level than as-sprayed Ti coatings (Figure 3f–j). The presence of considerable number of micro-pores and even (worse) micro-cracks in cold sprayed CP-Al coating microstructure (at inter-particle boundaries) on AZ91D Mg alloy was also reported by Y. Tao, et al. [22]. In fact, lower degree of localized plastic deformation (localized heating, stresses) [22] resulted in an extensive formation of micro-defects at inter-particle boundaries. Cold sprayed Al coating with high denseness and having sub-micron sized grains considerably improved the corrosion resistance compared to CP-Al bulk substrates [22,25].

Figure 4 shows the XRD spectra of bare Mg alloy, feedstock powders and as-cold sprayed coatings. XRD spectrum (Figure 4a) shows that bare AZ31B is mostly comprised of α-Mg phase. Powder particles and CS coatings displayed similar phase structure and crystal planes (Figure 4b,c). Phase transformation and oxidation weren't evidently observed in the CS coatings (Figure 4b,c). The broadened peaks in the XRD pattern of Ti (Figure 4c) coating are primarily related to the intense plastic deformation of powder particles in the coatings compared to Al coating (Figure 4b) during cold spray process [40,41]. Low processing temperature and sizable peening effect of the powder particles (during cold spray process) could lead to the retention of primary phase and crystal planes of powder particles in CS coatings. On the contrary, wire flamed sprayed Ti coatings were mainly constituted by oxides, nitrides, and carbides phases due to the nature of the flame spray process. This resulted in the inferior corrosion protection performance of the sprayed Ti coatings. Hence, these coatings had to be sealed with epoxy or Si resin for usage in the chloride containing solutions [31], so it is anticipated that a Ti coating (in this research) without any post-spray treatments could significantly decrease the corrosion rate of magnesium alloy and make AZ31B Mg alloy usable in chloride containing solutions for long periods of time.

Titanium coating considerably raised average micro-hardness ($HV_{0.025}$) of AZ31B Mg alloy surface (Figure 5a), while aluminum coating lowered average micro-hardness ($HV_{0.025}$) of substrate surface (Figure 5a). Higher micro-hardness in Ti coating may implies severe plastic deformation (high dislocation density) mostly at exterior region of powder particles (or inter-particle boundaries) that caused the increase in the coating denseness [13,25].

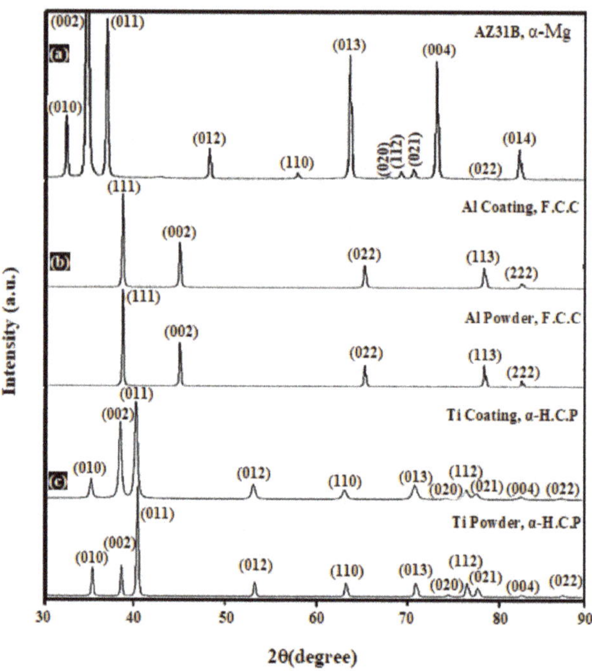

Figure 4. XRD patterns of (**a**) bare AZ31B, (**b**) as-received Al powder particles and as-sprayed Al coating, and (**c**) as-received Ti powder particles and as-sprayed Ti coating.

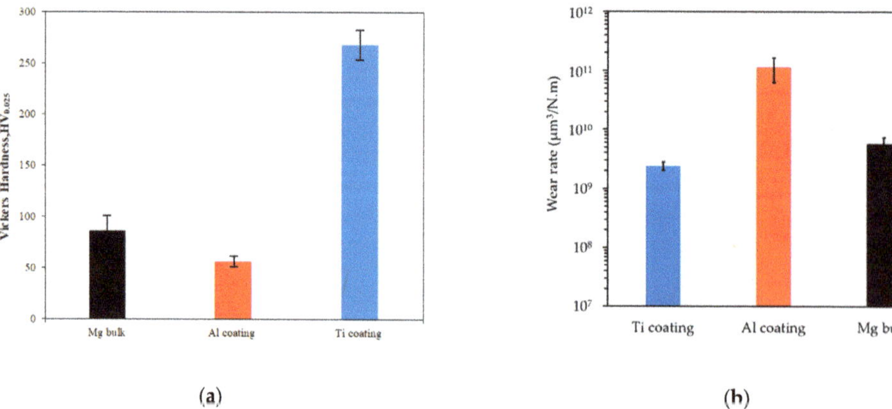

Figure 5. (**a**) Average micro-hardness ($HV_{0.025}$); and (**b**) wear rate of bare and coated AZ31B Mg substrates.

The wear depth on Ti-coated Mg alloy was around 7 μm in which Ti surface was in contact with the steel ball whereas the Al-coated Mg alloy displayed the highest wear depth of ~70 μm. The wear rate of the entire track calculated from the wear depth was plotted in Figure 5b. These results were also compared with bare AZ31BMg alloy samples that showed lower wear rate compared to Al coating surface but higher wear rate than Ti coating surfaces. In fact, a surface with higher hardness showed lower wear rate than a surface with lower hardness. This proves that CS Ti coating substantially raises surface hardness and lowers the wear rate of Mg alloys compared to CS Al coatings.

The mechanical and tribological characteristics of CS pure Ti coatings on Ti-6Al-4V substrates were studied by Khun et al. [36]. The results indicated that wear resistance of the CS Ti coating (experimented against steel balls) was noticeably higher than that of Ti-6Al-4V alloy. This was related to the cold work hardening (strain hardening) during spray process and interestingly highly wear-resistant oxide layers formation on wear tracks of CS Ti coatings (during wear tests). In fact, cold sprayed pure Ti coatings with higher compactness and lower porosities showed higher hardness and thus improved wear resistance on the Ti64 alloy as substrate [36]. Astarita, et al. also reported that CS Ti coatings can improve the wear performance of bare AA2024 alloy [42].

3.2. Electrochemical Behavior

3.2.1. Cyclic Potentiodynamic Polarization (CPP)

From the open circuit plots (Figure 6a) it can be seen that the surface of the samples was quite stable over the period of 1 h. The OCP values increase in the order: AZ31B < Al-coated AZ31B < Ti-coated AZ31B. It is evidently seen that bare Mg alloys with lower values of OCP are more susceptible to corrosion compared to coated Mg samples which showed higher values of OCP. But, Al-coated AZ31B is more active than Ti coated AZ31B.

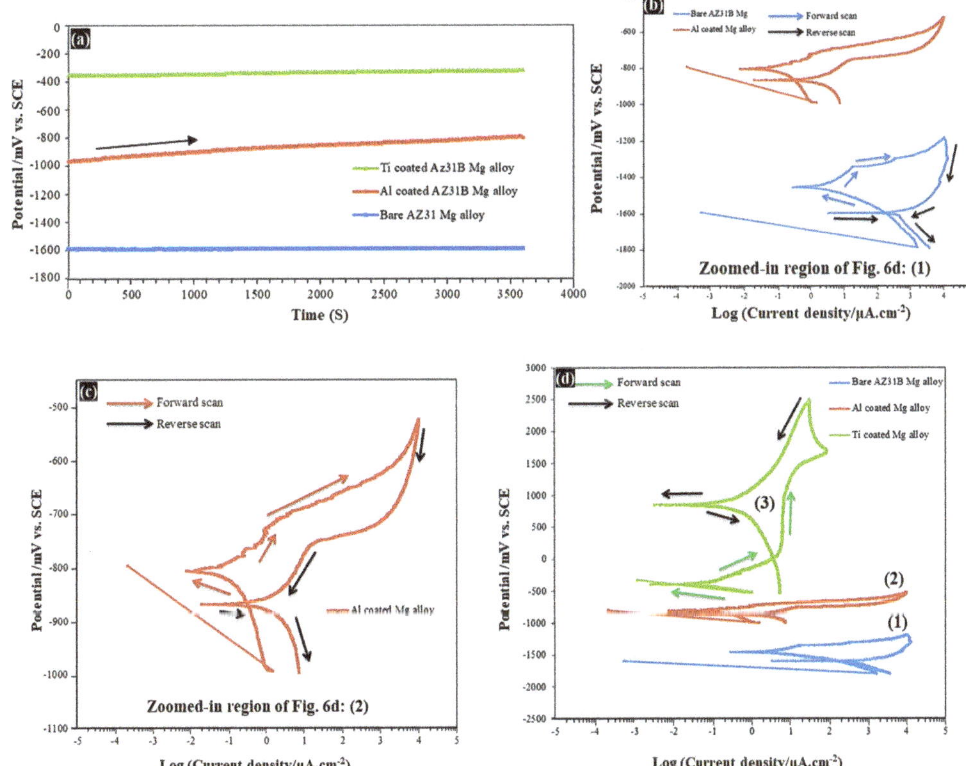

Figure 6. (**a**) Open circuit potential (OCP) for 1h, and (**b–d**) cyclic potentiodynamic polarization (CPP) curves for bare and coated AZ31B Mg alloys in 3.5 wt % NaCl electrolyte.

Cyclic potentiodynamic polarization tests (Figure 6b–d) were carried out to determine whether coated and uncoated AZ31B alloys experience pitting corrosion in chloride containing solutions. This test can also help whether the passive film formed on the surface has a tendency to heal itself (or repassivate) in the harsh environment. Moreover, CPP test was carried out three times to prove the repeatability of the obtained outcomes.

The corrosion parameters for the bare AZ31B samples as well as Al-and Ti-coated samples are given in Table 3. It is clear that Al-and Ti-coated AZ31B alloy samples have lower corrosion current densities (i_{corr}) compared to bare samples, where lowest i_{corr} and average corrosion rate ($P_i = 22.85 i_{corr}$ [43]) were observed for Ti-coated samples. The corrosion potential values were more noble for Al-coated and further higher for Ti-coated samples. Overall, it could be suggested that both Al and Ti cold sprayed coating improved the corrosion behavior of AZ31B alloy, obviously much better in case of Ti-coated samples. In this research, HPCS Ti coating considerably lowered i_{corr} to 0.049 $\mu A/cm^2$ from 2.504 $\mu A/cm^2$ and shifted E_{corr} to more noble potential, i.e., -387.299 mV_{SCE} from -1453.86 mV_{SCE} for AZ31B Mg alloy in 3.5 wt % NaCl solution. Nonetheless, magnetron sputtered Ti coating could only lower i_{corr} to 26.60 $\mu A/cm^2$ from 162.70 $\mu A/cm^2$ and E_{corr} to -1525 mV_{SCE} from -1570 mV_{SCE} for AZ91D Mg alloy in 3.5 wt % NaCl solution [32].

Table 3. The E_{corr}, and i_{corr}, β_a and β_c for bare and coated samples calculated from the forward scan of cyclic potentiodynamic polarization (CPP) curves.

Type of Sample	E_{corr} (mV_{SCE})	β_a (mV/dec)	β_c (mV/dec)	i_{corr} ($\mu A\ cm^{-2}$)
AZ31B	−1453.86	115.3	67.6	2.504
Al-coated AZ31B	−804.59	75.5	114	0.092
Ti-coated AZ31B	−387.299	159.9	99.7	0.049

The difference between pitting potential E_{pit} and E_{corr} ($E_{pit} - E_{corr}$) can be used as a measure of the propensity to the pitting nucleation [22]. Moreover, the difference between repassivation or protection potential (E_{rp}) and corrosion potential E_{corr} ($E_{rp} - E_{corr}$) can be employed as a measure of the repassivation ability. Lager values of ($E_{pit} - E_{corr}$) and ($E_{rp} - E_{corr}$) signify enhanced resistance to pitting corrosion and higher repassivation ability, respectively [22]. The $E_{rp} - E_{corr}$ values increase in the order: AZ31B < Al-coated AZ31B << Ti-coated AZ31B. On the reverse scan, AZ31B and Al-coated AZ31B showed a positive hysteresis loop, implying the further growth of pitting. It was reported that pitting corrosion can't get further expanded if reversed anodic curve shifted to lower current densities (as negative hysteresis loops) or the forward scan to be retraced by reversed curve. On the contrary, further pitting development is anticipated if reversed anodic curve shifted to higher current densities compared to forward scan (as positive hysteresis loops) [44]. The pits keep growing if $E_{corr} > E_{rp}$ and vice versa. Ti-coated Mg alloy indicates negative hysteresis loop, depicting repassivation of pits, in contrast to AZ31B and Al-coated AZ31B with positive hysteresis loops where $E_{corr} > E_{rp}$; indicating irreversible growth of pits. As-cold sprayed Nb coatings (from group 5B) also showed negative hysteresis loops. The repassivation behavior of CS Nb coating was attributed to the stored energy in the CS coatings assisting to passivate quickly and simply [45]. Analysis of $E_{pit} - E_{corr}$ values demonstrates that AZ31B and Al-coated AZ31B are most susceptible to pitting corrosion while Ti-coated Mg alloy indicates conspicuous resistance to pitting in chloride containing electrolyte.

The CPP tests also reveal that the anodic curves for AZ31B and Al-coated AZ31B had active current densities especially in case of AZ31B alloy. The cathodic current kinetics were highest for AZ31B compared to both Al-coated and Ti-coated alloys samples. The current density limit of 10 mA/cm^2 was reached at much lower potentials for AZ31B and Al-coated AZ31B alloy as well. Among all Ti-coated samples showed primary passivation followed by a secondary passivation (passive-like behavior) through a passive-active-passive (passive-like behavior) transition region. Interestingly, after the reaching the upper potential limit of 2.5 V_{SCE} the reverse scan showed lower current densities suggesting that the surface is still passive and lower potential cause lower dissolution for the passivated Ti surface. This is a very good indication of highly stable passive film which did not deteriorate even at such potentials.

3.2.2. Electrochemical Impedance Spectroscopy (EIS)

For further elucidation of the influence of cold sprayed coatings on the corrosion behavior of AZ31B Mg alloy, electrochemical impedance spectroscopy (EIS) of coated and uncoated Mg alloys was measured at their OCP. Nyquist plots are shown in Figure 7a,b. The impedance spectra was fitted by the electrical equivalent circuits (shown in Figure 7c,d: models 1 and 2). For the Nyquist diagram of uncoated sample (Figure 7d), one inductive loop and one capacitive loop at low frequency and high frequency were considered, respectively. This assumption is in harmony with the previous researches [26,46]. EEC (electrical equivalent circuit) for uncoated Mg alloy is comprised of R_{ct} (charge transfer resistance), R_s (electrolyte resistance), C_{dl} (CPE related to electrical double layer), adsorption inductance (L) and adsorption resistance (R_L) elements.

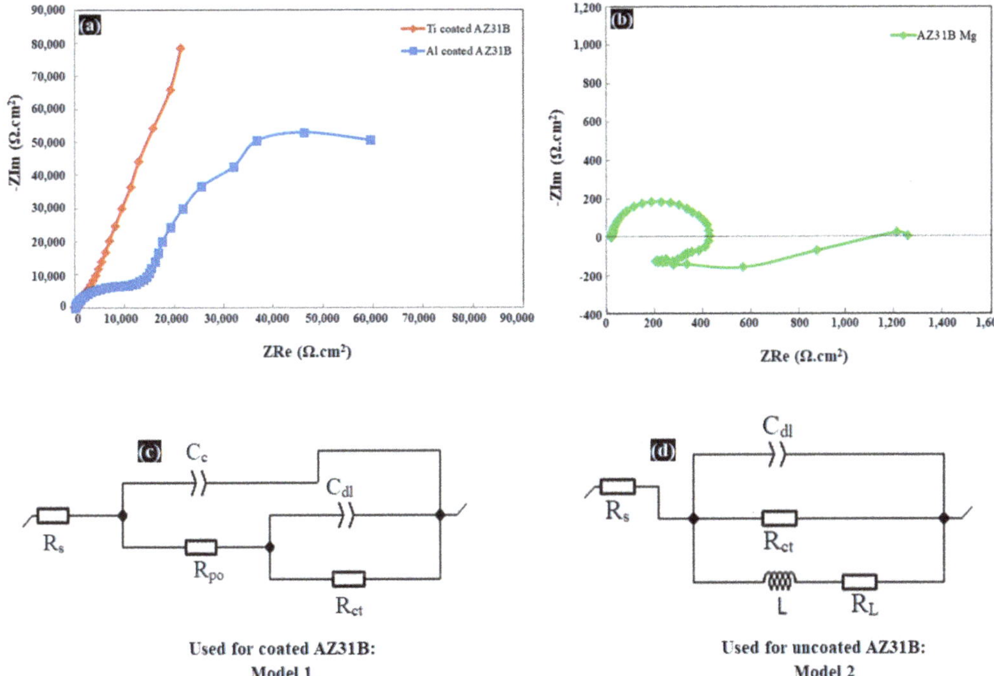

Figure 7. Nyquist plots of (**a**,**b**) coated and uncoated AZ31B Mg alloy substrates at OCP, electrical equivalent circuits to fit the impedance spectra of (**c**) coated AZ31B Mg alloys (model 1), and (**d**) bare AZ31B Mg alloys (model 2).

Impedance spectra of the coated samples were fitted by using the electrical equivalent circuit, as shown schematically in Figure 8. This schematically EEC includes R_s which is ohmic solution resistance at the working electrode/reference electrode interface; loop R_{ox}-C_{ox} which shows the resistance (R_c or R_{ox}) and capacitance (C_c or C_{ox}) of the oxide film; R_{ct} and C_{dl}; R_{po} that is the electrolyte resistance (as additional resistance) in the localized corrosion sites (and/or the pores). The R_c or R_{ox} values are pretty high and any conduction of electrons through the oxide layer has been reported to be impossible [47,48]. Thus, Mansfeld and Kendig [49] proposed the removal of this circuit element from EEC and replacement of EEC1 with EEC2 [47,48]. This simplified EEC2 has been also reported in the previous researches [49–52], so a capacitive loop at high frequency and another capacitive loop at low frequency regions were assumed for the coated Mg samples. Likewise, with regard to the non-ideality of the systems, the capacitors were replaced with the constant phase elements (CPE) for all samples [51,53].

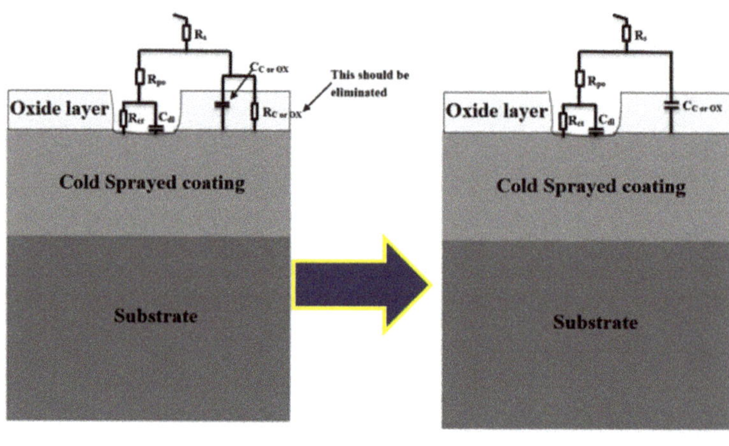

Figure 8. Schematic illustration of the developed EEC for fitting impedance spectra of CS Ti and CS Al-coated samples (per previous studies).

R_{ct} can predominantly control the electrochemical processes rate at the interface between electrode and electrolyte (or across the electrical double layer) [54,55]. As a matter of fact, corrosion rate is reversely proportional to the R_{ct} [56,57]. As presented in Table 4, Bare AZ31B Mg alloy with the lowest R_{ct} showed the maximum corrosion rate between 3 samples. This could be attributed to the low protective performance of the formed corrosion surface film on the Mg alloy surface in corrosive solution. Nevertheless, values of R_{ct} for Al- and Ti-coated samples were 164.707 kΩ·cm^2, and 5580 kΩ·cm^2, respectively. This indicates that Mg alloy could be protected by Al and Ti coatings. Nevertheless, R_{ct} value for the Ti-coated Mg alloy is roughly 34 times the value of R_{ct} for the Al-coated Mg alloy. The above-mentioned results reveal that the CS titanium coating can substantially declines the corrosion rate ($1/R_{ct}$) of Mg alloy in chloride containing solutions.

Table 4. EIS fitted results for bare and coated samples in 3.5 wt % NaCl electrolyte.

Samples	R_s (Ω cm^2)	R_{ct} (Ω cm^2)	Q_{dl} (F cm^{-2}s^{n-1})	n	R_{po} (Ω cm^2)	Q_c (F cm^{-2}s^{n-1})	n	R_L (Ω cm^2)	L (H cm^2)
Bare AZ31B Mg alloy	18.31	385.6	1.13×10^{-5}	0.98	-	-	-	187.9	187.5
Al-coated Mg alloy	15.18	164707	0.116×10^{-3}	0.87	14397	9.68×10^{-6}	0.88	-	-
Ti-coated Mg alloy	17.69	5.58×10^{6}	66.67×10^{-6}	0.798	6059	59.33×10^{-6}	0.90	-	-

It is worth mentioning that the corrosion resistance could be characterized by the parameter of polarization resistance at the corrosion potential (R_{pol}). Values of R_{pol} for Al and Ti-coated samples were calculated using Equation (2) under EEC2 (Model 1) in Figure 7c [58–63].

$$R_{pol} = R_{po} + R_{ct} \tag{2}$$

AZ31B coated with Ti coating showed the much better corrosion resistance than Al-coated Mg alloy in 3.5 wt % NaCl solution. It is expected that this unprecedented development could be maintained even during longer immersion times.

3.3. Long Term Immersion Test in 3.5 wt % NaCl Solution for 11 Days

The surface morphological characteristics of ground coated and uncoated AZ31BMg alloys after immersion test (in 3.5 wt % NaCl solution for 264 h) are shown in Figures 9–12. Uneven corrosion products with some micro-cracks [64] (Figure 9a,b,f) entirely covered the surface of the bare AZ31B Mg alloy surface. Corrosion products on the corroded bare AZ31B Mg alloy surface were mostly comprised of Mg, O, Cl as well as Al, and Na elements (Figure 9c–e). The presence of Mg, O and Cl elements (Figure 9d,e) in the corrosion products was related to the possible existence of $MgCl_2$ and $Mg(OH)_2$ phases [57]. Nonetheless, the lower corrosion current density, higher R_{ct}, and E_{corr} were all obtained for the AZ31BMg alloy when Al coating was applied on the AZ31B alloy (during short term electrochemical corrosion tests). Moreover, no significant corrosion was observed for the Al-coated AZ31B at the beginning of immersion test. However, deeper and even broad corrosion pits and also local net-cracks of corrosion products imply that the localized corrosion is worsened after 264 h of immersion (Figure 10a,b,f). In fact, the pitting density rises with uniform corrosion in the course of immersion time in chloride containing solutions. This behavior is also reported for the morphology of Al after corrosion [65,66].

Corrosion products on the corroded Al coating surface after 264 h of immersion were mainly constituted by Al, Mg and O as well as Cl and Na elements (Figure 10c–e). Mg element (from the AZ31B Mg substrate) was detected on the corroded Al coating surface. Al coating with low compactness could conduct chloride containing solution into the interior regions of the coating over immersion time. In fact, Al coating can't separate the AZ31B alloy surface from the corrosive electrolyte during long term immersion for 11 days.

High pressure cold sprayed Ti (from group 4B) coating (with high propensity to repassivation and also high pitting corrosion resistance) considerably mitigated the drawbacks associated with CP-Al coating (in this research). The corroded Ti coating surface didn't show any conspicuous corrosion pits and the other localized corrosions after long term immersion (Figure 11). It is interesting to note that, scratches (grinding tracks due to the coating surface preparation before the corrosion tests) are still distinguishable (Figure 11a,b,g) on the Ti coating surface even after 11 days of immersion.

Figure 11c–f show that corrosion products are primarily constituted by Ti and O elements which might correspond to existence of titanium-oxides [28] on the corroded coating surface. Likewise, Mg element (from AZ31B alloy substrate) wasn't detected on the corroded Ti layer surface. In fact, the Ti layer can considerably separate the AZ31BMg alloy surface from the corrosive electrolyte even during long-term immersion for 11 days.

On the contrary, galvanic cell formation between Mg alloy substrate and warm sprayed Ti coatings were observed by Moronczyk et.al [31]. This led to the corrosion products formation and their pile up at the interface between WS Ti coating and Mg alloy substrate. This finally caused the sudden rupture of the WS Ti coatings after merely 24 h of immersion test in 3.5 wt % NaCl electrolyte [31]. This behavior was also seen for magnetron sputtered Ti coating on AZ91D after 24 h of immersion in chloride coating solution [32]. This indicates that HPCS titanium coating not only modifies the hardness and wear behavior of Mg alloys, but also can exceptionally enhance the corrosion resistance of Mg alloys in aggressive solutions during long run immersion in chloride containing solution.

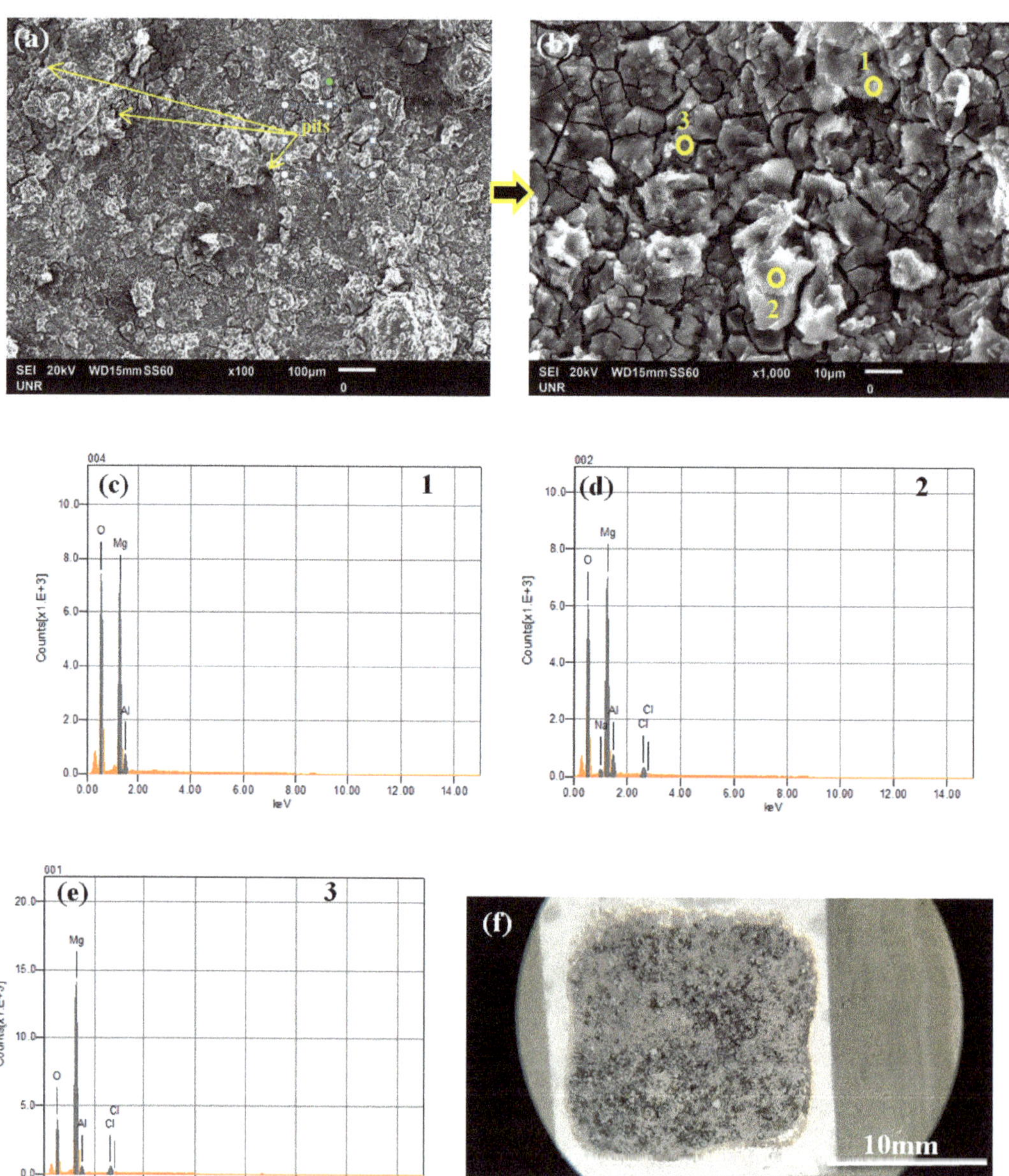

Figure 9. (**a**,**b**) surface morphology of ground surface of uncoated AZ31B Mg alloy after immersion test in 3.5 wt % NaCl solution for 11 days, EDS analysis of 1 (**c**), 2 (**d**), 3 (**e**) and (**f**) stereo microscope image at 50× magnification.

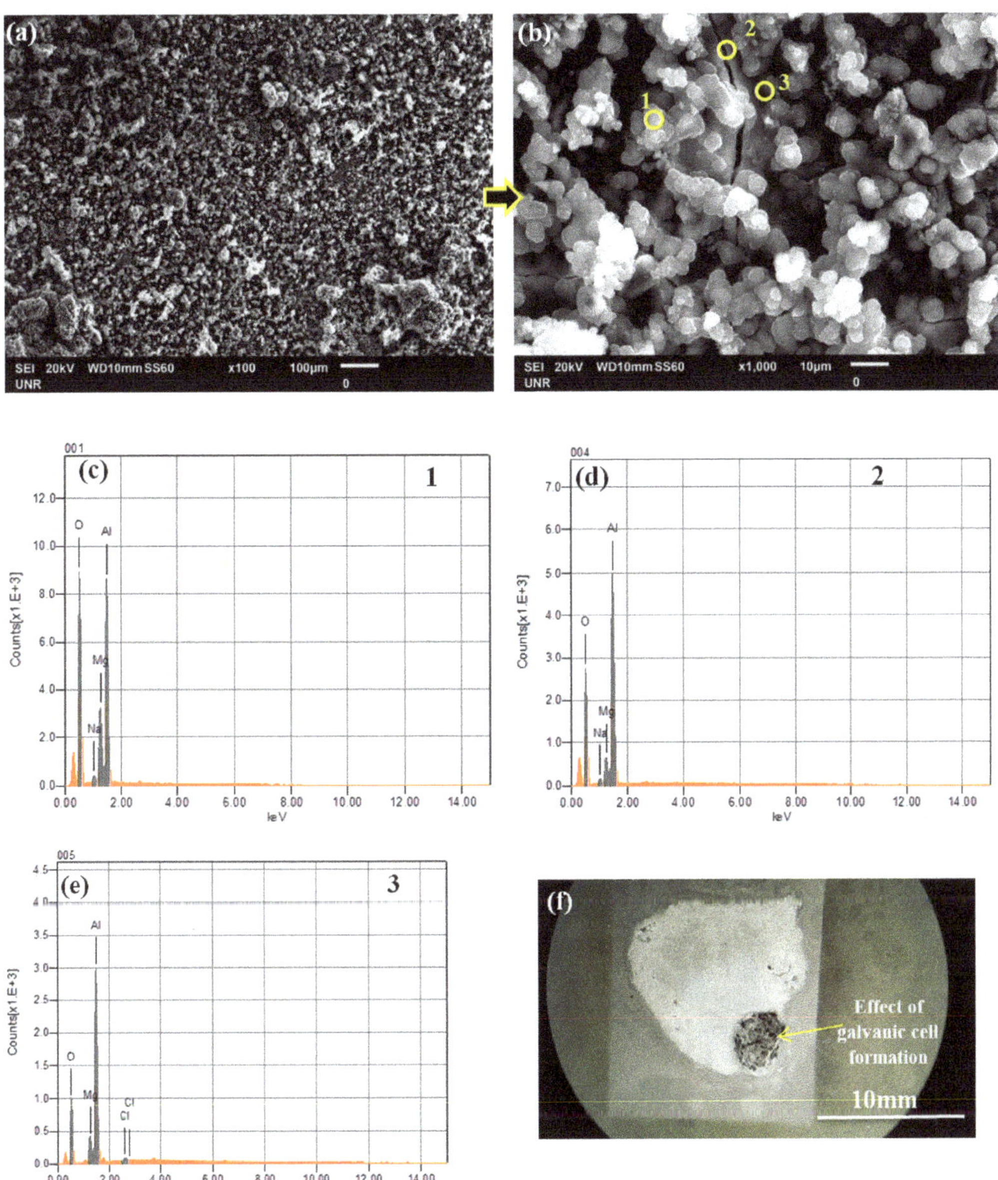

Figure 10. (**a**,**b**) surface morphology of ground surface of Al-coated AZ31B after immersion test in 3.5 wt % NaCl solution for 11 days, EDS analysis of 1 (**c**), 2 (**d**), 3 (**e**) and (**f**) stereo microscope image at 50× magnification.

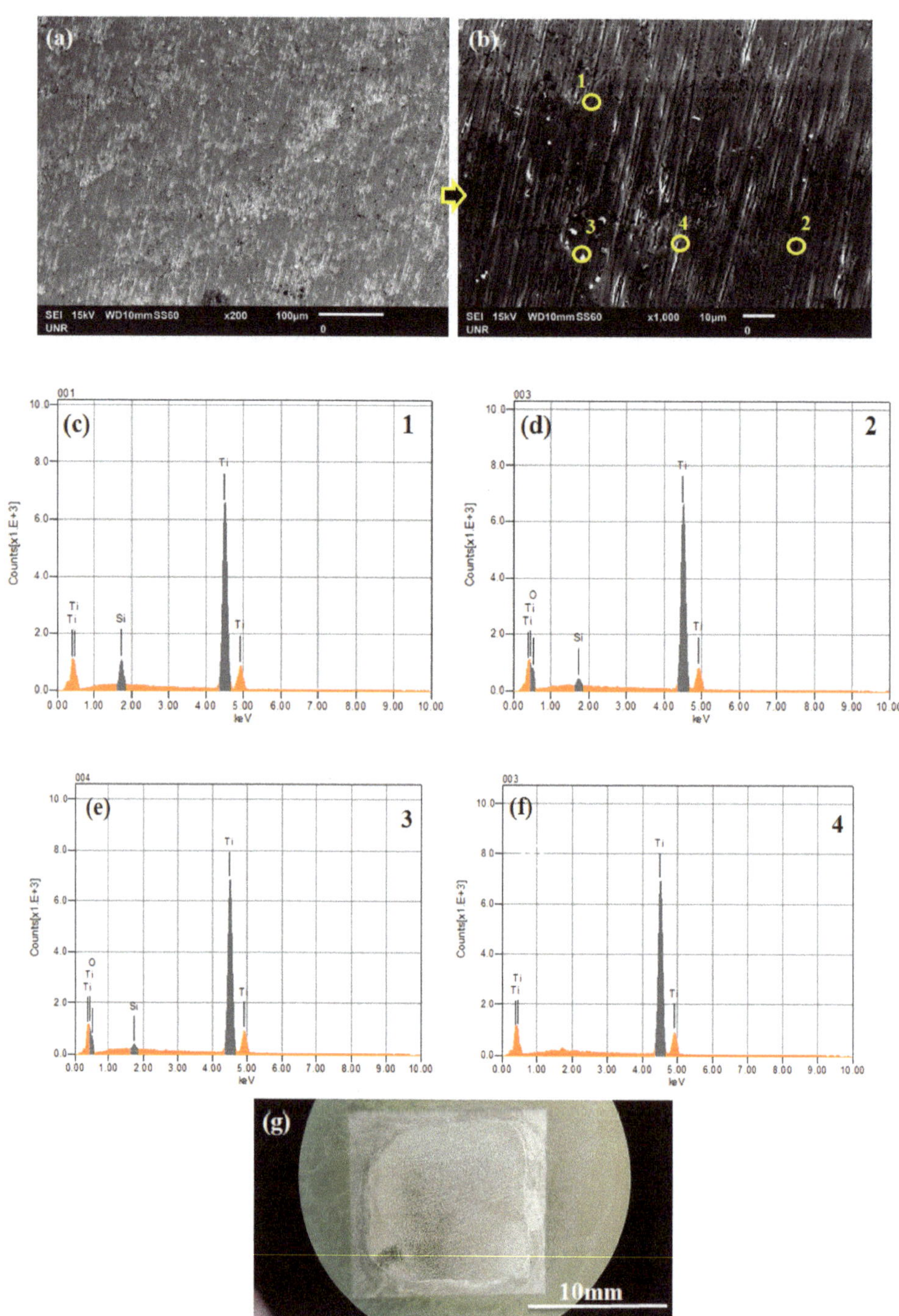

Figure 11. (**a**,**b**) surface morphology of ground surface of Ti-coated AZ31B after immersion test in 3.5 wt % NaCl solution for 11 days, EDS analysis of 1 (**c**), 2 (**d**), 3 (**e**), 4 (**f**) and (**g**) stereo microscope image at 50× magnification.

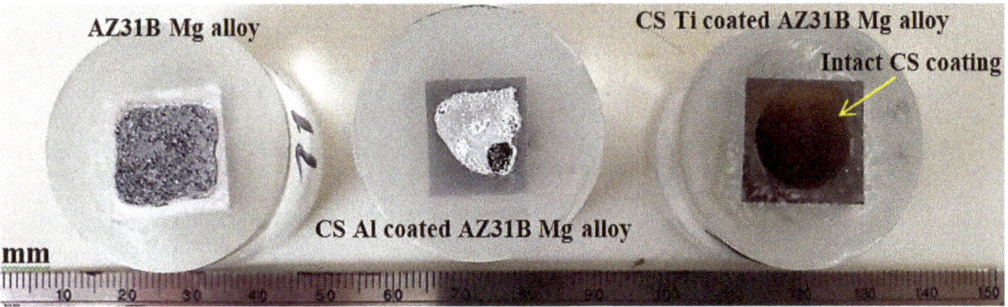

Figure 12. Photos of uncoated AZ31B Mg alloy, Al-coated AZ31B Mg alloy, and Ti-coated AZ31B Mg alloy (from left to right) after immersion test in 3.5 wt % NaCl solution for 11 days.

4. Conclusions

CS Ti coatings improved the hardness and wear behavior of Mg alloys compared to CS Al coatings. AZ31B coated with Ti coating showed the much better corrosion resistance than Al-coated Mg alloy in 3.5 wt % NaCl solution. AZ31B and Al-coated AZ31B were most vulnerable to pitting corrosion, while Ti-coated Mg alloy indicated extraordinary resistance to pitting in chloride containing electrolyte. Compared to CS Al coating, cold sprayed commercially pure-Ti coating exceptionally increased the repassivation ability of Mg alloys. Moreover, immersion test for 11 days further elucidated the effectiveness and corrosion protection performance of HPCS Ti coatings on Mg alloys in corrosive solutions.

Author Contributions: Conceptualization, M.D.; methodology, A.K.K., M.U.F.K. and C.M.K.; validation, A.K.K., M.U.F.K., P.L.M. and R.K.G.; formal analysis, M.D., A.K.K., M.U.F.K., P.L.M. and R.K.G.; investigation, M.D. and M.U.F.K.; resources, P.L.M., C.M.K., M.M. and R.K.G.; data curation, A.K.K., M.U.F.K., P.L.M. and R.K.G.; writing—original draft preparation, M.D., A.K.K. and M.U.F.K.; writing—review and editing, M.D., A.K.K., M.U.F.K., P.L.M., C.M.K., M.M. and R.K.G.; visualization, P.L.M.; supervision, M.D., P.L.M. and R.K.G.; project administration, M.D., C.M.K. and M.M.; funding acquisition, C.M.K., M.M. and R.K.G. All authors have read and agreed to the published version of the manuscript.

Funding: This research was funded by National Science Foundation, grant number NSF-CMMI 1846887.

Institutional Review Board Statement: Not applicable.

Informed Consent Statement: Not applicable.

Data Availability Statement: Data sharing not applicable.

Acknowledgments: The authors would like to acknowledge the ASB Industries, Inc for providing facilities and support. R. K. Gupta acknowledges the financial support from the National Science Foundation (NSF-CMMI 1846887) under the direction of Alexis Lewis. M. Misra duly acknowledges the startup fund from the University of Nevada: Reno.

Conflicts of Interest: The authors declare no conflict of interest.

References

1. Yi, A.; Du, J.; Wang, J.; Mu, S.; Zhang, G.; Li, W. Preparation and characterization of colored Ti/Zr conversion coating on AZ91D magnesium alloy. *Surf. Coat. Technol.* **2015**, *276*, 239–247. [CrossRef]
2. Daroonparvar, M.; Yajid, M.A.M.; Yusof, N.M.; Bakhsheshi-Rad, H.R.; Hamzah, E.; Kamali, H.A. Microstructural characterization and corrosion resistance evaluation of nanostructured Al and Al/AlCr coated Mg–Zn–Ce–La alloy. *Alloys Compd.* **2014**, *615*, 657–671. [CrossRef]
3. Singh, C.; Tiwari, S.K.; Singh, R. Exploring environment friendly nickel electrodeposition on AZ91 magnesium alloy: Effect of prior surface treatments and temperature of the bath on corrosion behavior. *Corros. Sci.* **2019**, *151*, 1–19. [CrossRef]

4. Daroonparvar, M.; Yajid, M.M.A.; Yusof, M.N.; Bakhsheshi-Rad, R.H. Fabrication and properties of triplex NiCrAlY/nano Al_2O_3·13% TiO_2/nano TiO_2 coatings on a magnesium alloy by atmospheric plasma spraying method. *Alloys Compd.* **2015**, *645*, 450–466. [CrossRef]
5. Carboneras, M.; López, M.D.; Rodrigo, P.; Campo, M.; Torres, B.; Otero, E.; Rams, J. Corrosion behaviour of thermally sprayed Al and Al/SiCp composite coatings on ZE41 magnesium alloy in chloride medium. *Corros. Sci.* **2010**, *52*, 761–768. [CrossRef]
6. Daroonparvar, M.; Yajid, M.M.; Bakhsheshi-Rad, R.H.; Yusof, M.N.; Izman, S.; Hamzah, E.; Kadir, A.M. Corrosion resistance investigation of nanostructured Si-and Si/TiO2-coated Mg alloy in 3.5% NaCl solution. *Vacuum* **2014**, *108*, 61–65.
7. Bakhsheshi-Rad, R.H.; Hamzah, E.; Ebrahimi-Kahrizsangi, R.; Daroonparvar, M.; Medraj, M. Fabrication and characterization of hydrophobic micro arc oxidation/poly-lactic acid duplex coating on biodegradable Mg–Ca alloy for corrosion protection. *Vacuum* **2016**, *125*, 185–188. [CrossRef]
8. Daroonparvar, M.; Yajid, M.M.; Bakhsheshi-Rad, H.R.; Ku-mar, P.; Kay, C.M.; Kalvala, P.R. Fabrication and Corrosion Resistance Evaluation of Novel Epoxy/Oxide Layer (MgO) Coating on Mg Alloy. *Prot. Met. Phys. Chem. Surf.* **2020**, *56*, 1039–1050. [CrossRef]
9. Daroonparvar, M.; Yajid, M.M.; Yusof, N.M.; Bakhsheshi-Rad, H.R.; Adabi, M.; Hamzah, E.; Kamali, H.A. Improvement of corrosion resistance of binary Mg-Ca alloys using duplex alu-minum-chromium coatings. *J. Mater. Eng. Perform.* **2015**, *7*, 2614–2627. [CrossRef]
10. Daroonparvar, M.; Yajid, M.M.; Gupta, R.K.; Yusof, N.M.; Bakhsheshi-Rad, H.R.; Ghandvar, H. Investigation of Corrosion Protection Performance of Multiphase PEO (Mg_2SiO_4, MgO, $MgAl_2O_4$) Coatings on Mg Alloy Formed in Aluminate-Silicate-based Mixture Electrolyte. *Prot. Met. Phys. Chem. Surf.* **2018**, *54*, 425–441. [CrossRef]
11. Widener, A.C.; Ellingsen, M.; Carter, M. Understanding Cold Spray for Enhanced Manufacturing Sustainability. *Mater. Sci. Forum* **2018**, *941*, 1867–1873. [CrossRef]
12. Monette, Z.; Kasar, K.A.; Daroonparvar, M.; Menezes, L.P. Supersonic particle deposition as an additive technology: Methods, challenges, and applications. *Int. J. Adv. Manuf. Technol.* **2020**, *106*, 2079–2099. [CrossRef]
13. Kay, C.M.; Karthikeyan, J. *High Pressure Cold Spray: Principles and Applications*; ASM Thermal Spray Society: Materials Park, OH, USA, 2016; ISBN 978-1-62708-096-5.
14. Zulkifli, I.S.M.; Yajid, M.A.M.; Idris, M.H.; Uday, M.B.; Daroonparvar, M.; Emadzadeh, A.; Arshad, A. Micro-structural evaluation and thermal oxidation behaviors of YSZ/NiCoCrAlYTa coatings deposited by different thermal techniques. *Ceram. Int.* **2020**, *46*, 22438–22451. [CrossRef]
15. Daroonparvar, M.; Yajid, M.A.M.; Kay, C.M.M.; Bakhsheshi-Rad, H.; Gupta, R.K.; Yusof, N.M.; Ghandvar, H.; Arshad, A.; Zulkifli, I.S.M. Effects of Al_2O_3 diffusion barrier layer (including Y-containing small oxide precipitates) and nanostructured YSZ top coat on the oxidation behavior of HVOF NiCoCrAl-TaY/APS YSZ coatings at 1100 °C. *Corros. Sci.* **2018**, *144*, 13–34. [CrossRef]
16. Luo, X.T.; Li, C.X.; Shang, F.L.; Yang, G.J.; Wang, Y.Y.; Li, C.J. High velocity impact induced microstructure evolution during deposition of cold spray coatings. *Surf. Coat. Technol.* **2014**, *254*, 11–20. [CrossRef]
17. Assadi, H.; Kreye, H.; Gärtner, F.; Klassen, T. Cold spraying: A materials perspective. *Acta Mater.* **2016**, *116*, 382–407. [CrossRef]
18. Daroonparvar, M.; Yajid, M.A.M.; Yusof, N.M.; Bakhsheshi-Rad, H.R.; Hamzah, E. Microstructural characterisation of air plasma sprayed nanostructure ceramic coatings on Mg–1% Ca alloys (bonded by NiCoCrAlYTa). *Ceram. Int.* **2016**, *42*, 357–371. [CrossRef]
19. Daroonparvar, M.; Yajid, M.M.; Yusof, N.M.; Bakhsheshi-Rad, H.R.; Hussain, M.; Hamzah, E. Evaluation of Normal and Nanolayer Composite Thermal Barrier Coatings in Fused Vanadate-Sulfate Salts at 1000 °C. *Adv. Mater. Sci. Eng.* **2013**, *2013*, 1–13. [CrossRef]
20. Bakhsheshi-Rad, H.R.; Abdellahi, M.; Hamzah, E.; Daroonparvar, M.; Rafiei, M. Introducing a composite coating containing CNTs with good corrosion properties: Characterization and simulation. *RSC Adv.* **2016**, *6*, 108498–108512. [CrossRef]
21. Wang, J.; Pang, X.; Jahed, H. Surface protection of Mg alloys in automotive applications: A review. *Mater. Sci.* **2019**, *6*, 567–600. [CrossRef]
22. Tao, Y.; Xiong, T.; Sun, C.; Kong, L.; Cui, X.; Li, T.; Song, G.L. Microstructure and corrosion performance of a cold sprayed aluminium coating on AZ91D magnesium alloy. *Corros. Sci.* **2010**, *52*, 3191–3197. [CrossRef]
23. DeForce, B.S.; Eden, T.J.; Potter, J.K. Cold spray Al–5% Mg coatings for the corrosion protection of magnesium alloys. *J. Therm. Spray Technol.* **2011**, *20*, 1352–1358. [CrossRef]
24. Diab, M.; Pang, X.; Jahed, H. The effect of pure aluminum cold spray coating on corrosion and corrosion fatigue of magnesium (3% Al–1% Zn) extrusion. *Surf. Coat. Technol.* **2017**, *309*, 423–435. [CrossRef]
25. Ngai, S.; Ngai, T.; Vogel, F.; Story, W.; Thompson, G.B.; Brewer, L.N. Saltwater corrosion behavior of cold sprayed AA7075 aluminum alloy coatings. *Corros. Sci.* **2018**, *130*, 231–240. [CrossRef]
26. Wei, Y.K.; Li, Y.J.; Zhang, Y.; Luo, X.T.; Li, C.J. Corrosion resistant nickel coating with strong adhesion on AZ31B magnesium alloy prepared by an in-situ shot-peening-assisted cold spray. *Corros. Sci.* **2018**, *138*, 105–115. [CrossRef]
27. Zhang, Z.; Liu, F.; Han, E.H.; Xu, L. Mechanical and corrosion properties in 3.5% NaCl solution of cold sprayed Al-based coatings. *Surf. Coat. Technol.* **2020**, *385*, 125372. [CrossRef]
28. Prando, D.; Brenna, A.; Diamanti, M.V.; Beretta, S.; Bolzoni, F.; Ormellese, M.; Pedeferri, M. Corrosion of titanium. Part 1: Aggressive environments and main forms of degradation. *J. Appl. Biomater. Funct. Mater.* **2017**, *15*, 291–302. [CrossRef]
29. Sun, J.; Han, Y.; Cui, K. Innovative fabrication of porous titanium coating on titanium by cold spraying and vacuum sintering. *Mater. Lett.* **2008**, *62*, 3623–3625. [CrossRef]

30. Li, W.; Cao, C.; Yin, S. Solid-state cold spraying of Ti and its alloys: A literature review. *Prog. Mater. Sci.* **2020**, *110*, 100633. [CrossRef]
31. Morończyk, B.; Ura-Bińczyk, E.; Kuroda, S.; Jaroszewicz, J.; Molak, R.M. Microstructure and corrosion resistance of warm sprayed titanium coatings with polymer sealing for corrosion protection of AZ91E magnesium alloy. *Surf. Coat. Technol.* **2019**, *363*, 142–151. [CrossRef]
32. Zhang, D.; Wei, B.; Wu, Z.; Qi, Z.; Wang, Z. A comparative study on the corrosion behavior of Al, Ti, Zr and Hf metallic coatings deposited on AZ91D magnesium. *Surf. Coat. Technol.* **2016**, *303*, 94–102. [CrossRef]
33. *ASTM E2109-01 Standard Test Methods for Determining Area Percentage Porosity in Thermal Sprayed Coatings*; ASTM: West Conshohocken, PA, USA, 2014.
34. *ASTM G133-05 Standard Test Method for Linearly Reciprocating Ball-on-Flat Sliding Wear*; ASTM: West Conshohocken, PA, USA, 2005.
35. Menezes, P.L.; Nosonovsky, M.; Ingole, S.P.; Kailas, S.V.; Lovell, M.R. *Tribology for Scientists and Engineers*; Springer: Berlin, Germany, 2013.
36. Khun, N.W.; Tan AW, Y.; Liu, E. Mechanical and tribological properties of cold sprayed Ti coatings on Ti-6Al-4V substrates. *J. Therm. Spray Technol.* **2016**, *25*, 715–724. [CrossRef]
37. Siddique, S.; Li, C.X.; Bernussi, A.A.; Hussain, S.W.; Yasir, M. Enhanced Electrochemical and Tribo-logical Properties of AZ91D Mg Alloy via Cold Spraying of Al Alloy. *J. Therm. Spray Technol.* **2019**, *28*, 1739–1748. [CrossRef]
38. *ASTM Standard G61-86 Standard Test Method for Conducting Cyclic Potentiodynamic Polarization (CPP) Measurements for Localized Corrosion Susceptibility of Iron-, Nickel-, or Cobalt-Based Alloys*; ASTM: West Conshohocken, PA, USA, 2014.
39. Cizek, J.; Kovarik, O.; Siegl, J.; Khor, K.A.; Dlouhy, I. Influence of plasma and cold spray deposited titanium Layers on high-cycle fatigue properties of Ti6Al4V substrates. *Surf. Coat. Technol.* **2013**, *217*, 23–33. [CrossRef]
40. Sun, W.; Tan, A.W.Y.; Marinescu, I.; Toh, W.Q.; Liu, E. Adhesion, tribological and corrosion properties of cold-sprayed CoCrMo and Ti6Al4V coatings on 6061-T651 Al alloy. *Surf. Coat. Technol.* **2017**, *326*, 291–298. [CrossRef]
41. Sun, W.; Tan, A.W.Y.; Khun, N.W.; Marinescu, I.; Liu, E. Effect of substrate surface condition on fatigue behavior of cold sprayed Ti6Al4V coatings. *Surf. Coat. Technol.* **2017**, *320*, 452–457. [CrossRef]
42. Astarita, A.; Rubino, F.; Carlone, P.; Ruggiero, A.; Leone, C.; Genna, S.; Merola, M.; Squillace, A. On the Improvement of AA2024 Wear Properties through the Deposition of a Cold-Sprayed Titanium Coating. *Metals* **2016**, *6*, 185. [CrossRef]
43. Shi, Z.; Liu, M.; Atrens, A. Measurement of the corrosion rate of magnesium alloys using Tafel extrapolation. *Corros. Sci.* **2010**, *52*, 579–588. [CrossRef]
44. Liu, Y.; Meng, G.Z.; Cheng, Y.F. Electronic structure and pitting behavior of 3003 aluminum alloy passivated under various conditions. *Electrochim. Acta* **2009**, *54*, 4155–4163. [CrossRef]
45. Kumar, S.; Jyothirmayi, A.; Wasekar, N.; Joshi, S.V. Influence of annealing on mechanical and electrochemical properties of cold sprayed niobium coatings. *Surf. Coat. Technol.* **2016**, *296*, 124–135. [CrossRef]
46. Ezhilselvi, V.; Nithin, J.; Balaraju, J.N.; Subramanian, S. The influence of current density on the morphology and corrosion properties of MAO coatings on AZ31B magnesium alloy. *Surf. Coat. Technol.* **2016**, *288*, 221–229. [CrossRef]
47. Moreto, J.A.; Marino, C.E.B.; Bose Filho, W.W.; Rocha, L.A.; Fernandes, J.C.S. SVET, SKP and EIS study of the corrosion behavior of high strength Al and Al–Li alloys used in aircraft fabrication. *Corros. Sci.* **2014**, *84*, 30–41. [CrossRef]
48. Kendig, M.W.; Mansfeld, F. AC electrochemical impedance of a model pit. *J. Electrochem. Soc.* **1982**, *129*, C318.
49. Guo, F.; Jiang, W.; Tang, G.; Xie, Z.; Dai, H.; Wang, E.; Chen, Y.; Liu, L. Enhancing anti-wear and anti-corrosion performance of cold spraying aluminum coating by high current pulsed electron beam irradiation. *Vacuum* **2020**, *182*, 109772. [CrossRef]
50. Da Silva, F.S.; Bedoya, J.; Dosta, S.; Cinca, N.; Cano, I.G.; Guilemany, J.M.; Benedetti, A.V. Corrosion characteristics of cold gas spray coatings of reinforced aluminum deposited onto carbon steel. *Corros. Sci.* **2017**, *114*, 57–71. [CrossRef]
51. López-Ortega, A.; Arana, J.L.; Rodríguez, E.; Bayón, R. Corrosion, wear and tribo-corrosion performance of a thermally sprayed aluminum coating modified by plasma electrolytic oxidation technique for offshore submerged components protection. *Corros. Sci.* **2018**, *143*, 258–280. [CrossRef]
52. Lu, F.F.; Ma, K.; Li, C.X.; Yasir, M.; Luo, X.T.; Li, C.J. Enhanced corrosion resistance of cold-sprayed and shot-peened aluminum coatings on LA43M magnesium alloy. *Surf. Coat. Technol.* **2020**, *394*, 125865. [CrossRef]
53. Shi, Z.; Song, G.; Cao, C.N.; Lin, H.; Lu, M. Electrochemical potential noise of 321 stainless steel stressed under constant strain rate testing conditions. *Electrochim. Acta* **2007**, *52*, 2123–2133. [CrossRef]
54. Chen, L.; Gu, Y.; Liu, L.; Liu, S.; Hou, B.; Liu, Q.; Ding, H. Effect of ultrasonic cold forging technology as the pretreatment on the corrosion resistance of MAO Ca/P coating on AZ31B Mg alloy. *J. Alloys Compd.* **2015**, *635*, 278–288.
55. Liu, M.; Mao, X.; Zhu, H.; Lin, A.; Wang, D. Water and corrosion resistance of epoxy–acrylic–amine waterborne coatings: Effects of resin molecular weight, polar group and hydrophobic segment. *Corros. Sci.* **2013**, *75*, 106–113.
56. Daroonparvar, M.; Yajid, M.A.M.; Yusof, N.M.; Bakhsheshi-Rad, H.R. Preparation and corrosion resistance of a nanocomposite plasma electrolytic oxidation coating on Mg–1%Ca alloy formed in aluminate electrolyte containing titania nano-additives. *Alloys Compd.* **2016**, *68*, 841–857. [CrossRef]
57. Daroonparvar, M.; Yajid, M.A.M.; Yusof, N.M.; Bakhsheshi-Rad, H.R.; Hamzah, E.; Mardanikivi, T. Deposition of duplex MAO layer/nanostructured titanium dioxide composite coatings on Mg–1% Ca alloy using a combined technique of air plasma spraying and micro arc oxidation. *Alloys Compd.* **2015**, *649*, 591–605. [CrossRef]

58. Bordbar-Khiabani, A.; Yarmand, B.; Sharifi-Asl, S.; Mozafari, M. Improved corrosion performance of biodegradable magnesium in simulated inflammatory condition via drug-loaded plasma electrolytic oxidation coatings. *Mater. Chem. Phys.* **2020**, *239*, 122003. [CrossRef]
59. Hu, J.; Li, Q.; Zhong, X.; Zhang, L.; Chen, B. Composite anticorrosion coatings for AZ91D magnesium alloy with molybdate conversion coating and silicon sol–gel coatings. *Prog. Org. Coat.* **2009**, *66*, 199–205. [CrossRef]
60. Sowa, M.; Simka, W. Electrochemical Impedance and Polarization Corrosion Studies of Tantalum Surface Modified by DC Plasma Electrolytic Oxidation. *Materials* **2018**, *11*, 545. [CrossRef]
61. Wang, C.; Fang, H.; Hang, C.; Sun, Y.; Peng, Z.; Wei, W.; Wang, Y. Fabrication and characterization of silk fibroin coating on APTES pretreated Mg-Zn-Ca alloy. *Mater. Sci. Eng. C* **2020**, *110*, 110742. [CrossRef]
62. Merl, D.K.; Panjan, P.; Kovač, J. Corrosion and surface study of sputtered Al–W coatings with a range of tungsten contents. *Corros. Sci.* **2013**, *69*, 359–368. [CrossRef]
63. Daroonparvar, M.; Khan, M.U.F.; Saadeh, Y.; Kay, C.M.; Kasar, A.K.; Kumar, P.; Esteves, L.; Misra, M.; Menezes, P.; Kalvala, P.R. Modification of surface hardness, wear re-sistance and corrosion resistance of cold spray Al coated AZ31B Mg alloy using cold spray double layered Ta/Ti coating in 3.5wt % NaCl solution. *Corros. Sci.* **2020**, *176*, 109029. [CrossRef]
64. Cui, X.J.; Lin, X.Z.; Liu, C.H.; Yang, R.S.; Zheng, X.W.; Gong, M. Fabrication corrosion resistance of a hydrophobic micro- arc oxidation coating on AZ31 Mg alloy. *Corros. Sci.* **2015**, *90*, 402–412. [CrossRef]
65. Farooq, A.; Hamza, M.; Ahmed, Q.; Deen, K.M. Evaluating the performance of zinc and aluminum sacrificial anodes in artificial seawater. *Electrochim. Acta* **2019**, *314*, 135–141. [CrossRef]
66. Cui, L.Y.; Gao, S.D.; Li, P.P.; Zeng, R.C.; Zhang, F.; Li, S.Q.; Han, E.H. Corrosion resistance of a self-healing micro-arc oxidation/polymethyltrimethoxysilane composite coating on magnesium alloy AZ31. *Corros. Sci.* **2017**, *118*, 84–95. [CrossRef]

Article

Molecular Dynamics Simulation of Dislocation Plasticity Mechanism of Nanoscale Ductile Materials in the Cold Gas Dynamic Spray Process

Sunday Temitope Oyinbo * and Tien-Chien Jen *

Department of Mechanical Engineering Science, University of Johannesburg, Gauteng 2006, South Africa
* Correspondence: soyinbo@uj.ac.za (S.T.O.); tjen@uj.ac.za (T.-C.J.)

Received: 21 October 2020; Accepted: 3 November 2020; Published: 10 November 2020

Abstract: The dislocation plasticity of ductile materials in a dynamic process of cold gas spraying is a relatively new research topic. This paper offers an insight into the microstructure and dislocation mechanism of the coating using simulations of molecular dynamics (MD) because of the short MD simulation time scales. The nano-scale deposition of ductile materials onto a deformable copper substrate has been investigated in accordance with the material combination and impact velocities in the particle/substrate interfacial region. To examine the jetting mechanisms in a range of process parameters, rigorous analyses of the developments in pressure, temperature, dislocation plasticity, and microstructure are investigated. The pressure wave propagation's critical function was identified by the molecular dynamics' simulations in particle jet initiation, i.e., exterior material flow to the periphery of the particle and substrate interface. The initiation of jet occurs at the point of shock waves interact with the particle/substrate periphery and leads to localization of the metal softening in this region. In particular, our findings indicate that the initial particle velocity significantly influences the interactions between the material particles and the substrate surface, yielding various atomic strain and temperature distribution, processes of microstructure evolution, and the development of dislocation density in the particle/substrate interfacial zone for particles with various impact velocities. The dislocation density in the particle/substrate interface area is observed to grow much more quickly during the impact phase of Ni and Cu particles and the evolution of the microstructure for particles at varying initial impact velocities is very different.

Keywords: ductile materials; dislocation density; microstructure and recrystallization; cold gas dynamic spray; molecular dynamics

1. Introduction

Cold gas dynamic spraying (CGDS) is a modern additive manufacturing approach and a promising technique in the field of materials processing that recently has been implemented for several industrial applications. The CGDS process is mainly a powder deposition process, which uses the ability to self-consolidate the solid particles that bond together at their solid-state. Such strong bonding self-consolidation ability resulted from a high-velocity (supersonic-velocity) impact [1–7]. Thurston developed this technique at the beginning of the twentieth century [8]. Afterwards, a pressurized or blast gas to propel metal powders up to a maximum speed approximately 300 m/s and then create a deposit via a high-velocity collision with a substrate. In the 1950s, the modern Rocheville technology with the gas flow across the De-Laval nozzle was an important breakthrough that, at that time, allowed the speeds to be higher than those of current methodologies and created a consistent thin coating. The cold gas dynamic spray process phenomenological behaviour was studied by the Russian Academy, Institute of Applied and Theoretical Mechanics further in the 1980s [8]. Their discoveries resulted in the creation of novel patents of cold gas-dynamic spraying devices and experimental processes for the production of cold gas dynamic sprays which eventually resulted in reliable additive manufacturing processes. While several viability studies show the feasibility of cold gas dynamic spray, deposit development and bonding mechanisms are continuously being studied to extend the materials concerned.

Two phases, including particle/substrate adhesion and the deposit growth, rule the deposition process during CGDS. The distinct phenomena of bonding mechanisms characterize every phase. Interparticle cohesion is recommended for ductile materials (metals in terms of deposit growth due to plastic deformation) such as copper (Cu), aluminum (Al), nickel (Ni), and silver (Ag), etc. The cohesive effect of interfaces is considered to occur through atomic interactions, because of intimate metallurgical interaction during the transformation of the phase, while the interfacial zone is subjected to high impact collision and experiences an extreme plastic deformation rate [9–12]. By comparison, it was also possible to recognize the self-compaction, fragmentation, and the final deposit consolidation due to the interlocking and stacking of fragments, particularly for non-ductile materials e.g., ceramics.

Researchers have demonstrated numerical analysis and experimental findings of the bonding mechanisms, which primarily occur by mechanical anchoring, metallurgical bonding, interfacial mixing, or mechanical locking, in literature for CGDS. Metallurgical bonding is possible because of a dynamic recrystallization phenomenon [13], a hyper-quenching phenomenon caused by a substantially large plastic strain (adiabatic shearing) in the interface of interfacial confinement and the formation of an amorphous middle layer covering intermetallic region [14]. Mechanical anchoring is instigated by the slight indentation on the substrate by the particles, which ensures that the particles are anchored, and which is primarily seen in combinations of the metallic component with the substrates made of ceramic [15–17]. Mechanical interlocking means particle/substrate integration as a result of in-depth penetration of particle into the substrates combinations as follows: metal/metal [18,19], ceramic/metal [20], oxide/polymer [21], and metal/polymer [22]. In the case of mechanical particle deformation within a geometrical imperfection of the substrate surface, the concept of inter-locking can also be applied [23,24]. This is also an understanding of the material consistency through the surface, produced during the soft particle deposition on a hard substrate. Such instances of these occurrences are soft polymer/metal [25,26], metal/ceramic [17] and metal/polymer [1,23,24,27]. The adhesion mechanism also regulates the production of interfacial vortices during interfacial mixing which permits the intermix of particles and substrate across the interface [24,28,29]

Since a broad range of new and progressive materials can be deposited with CGDS, academics and industry are increasingly interested in the CGDS technology. The CGDS approach provides different functional features for several obtainable industrial applications and significant progress is also anticipated in the coming decades. Several deposits of material can now be achieved [30,31]. They can be categorized according to their deposition technique and materials type. This comprises three distinct categories: (1) single material deposits, (2) a mixture of different particles, composite-based deposits, and (3) a nanomaterials deposit (i.e., a deposit creating nanosized characteristics). Additionally, the adhesion mechanisms versatility of the CGDS method suggests an additional category of the deposit as material hybridization among particles and substrates. The specific form of deposit also takes into account the possibility of hybridization and is called "hybrid particle/substrate assembly."

The study into the processes of cold gas dynamic spray mechanism of surface generation focuses primarily on the peening effect [32], localized softening [33], pressure waves [34], recrystallization [35], size effect [36,37], localization deformation [38,39], bonding [28,40], adhesive strength [41–43], oxide destruction [44], crystal orientation effect [36,45], evolution of microstructure [46,47] in stress/strain and nanoindentation [48], and so on. The result of the material combination and initial impact velocity on the dislocation plasticity and cold gas dynamic spray coating surface microstructural processes is barely explored. Since deformation of particles takes place within a quite short duration of 10^{-9} s order, it is very complicated to find the answers in-situ investigations [49]. Generally, detailed observations of the experimental study in a CGDS can usually be performed only after the spraying is completed and typically relies on particle cross-section microscopy observations in the as-deposited condition [50–52]. Simulations of molecular dynamics (MD) provide us with an effective means of controlling and examining the complex nanoscale atomically structure and behaviour [53].

Cu, Al, Ni, and Ag, a ductile face-centered cubic transition metal, is prototypical in both CGDS multiscale models and particle deposition experiments. In this paper, we present atomic research with MD on the single supersonic-speed impact of Al, Cu, Ni, and Ag particles on a copper substrate. To describe at the fundamental level the deformation process of ductile material particles, our research centered on the basic mechanisms of metal hardening and dislocation plasticity. The mechanisms of plastic deformation are studied from the perspective of the evolution of atomic structure through jet initiation, dislocation density, and the evolution of microstructural transformation. The dislocation density at the interface of particle/substrate interfacial region is observed to grow much more quickly during the impact phase of Ni and Cu particles and the evolution of the microstructure for particles at varying initial impact speeds is very different.

2. Computational Approach

The simulations of molecular dynamics (MD) were used to analyze the deposition behaviour of ductile nanoscale material particles in compliance with material combination and impact velocities. The large-scale atomic/molecular massively parallel simulator (LAMMPS) package [54] has been used to conduct MD simulations of nano-scale particles impaction. To understand the deformation process during particle deposition, OVITO [55], an accessible visualization tool, was used to examine the cross-section of the impact region. Figure 1 displays a diagram of the preliminary 3D simulation model for the deposition of nanoscale particles onto a Cu substrate. The analysis of the deposition characteristic of nanoscale ductile material preparation was considered in deformable spherical nano-scale Al, Ni, Cu, and Ag particles of 400 Å diameter and impact velocities of 500 to 1500 m/s. The particle material consists of around 2,051,820 atoms. The substrate is aligned in the [1 0 0], [0 1 0] and [0 0 1] for x-, y- and z- crystallographic direction, respectively. The substrate material consists of a face-centered cubic (FCC), with 3.61 Å lattice constant. The Cu substrate dimensions along an x-, y-, and z-direction are 700 Å × 700 Å × 600 Å. The initial distance from the particles to the substrate surface is 20 Å. The x-, y-, and z-directions are subjected to periodic boundary conditions (p p p). The substrate consists of the fixed boundary layer (700 Å × 700 Å × 50 Å), the thermostat layer (700 Å × 700 Å × 100 Å) and dynamic layer (700 Å × 700 Å × 500 Å. The phase time is 1.0 fs.

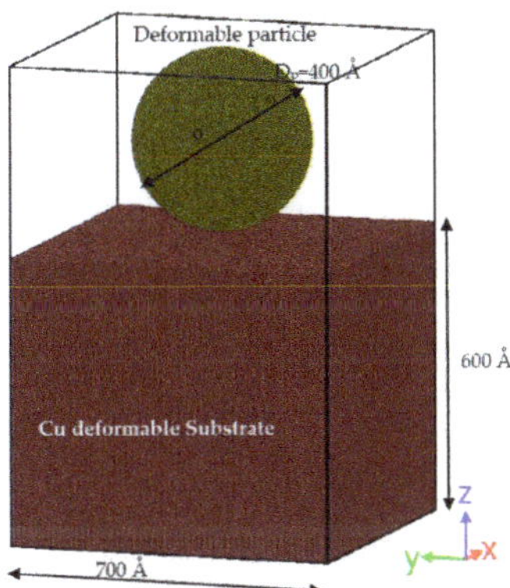

Figure 1. Cold gas dynamic spray MD simulation model.

In the setup of the first simulation, the thermal region maintained at 273 K was set into contact with the system via thermostat layer. This thermally linked area (the "heat sink") was the channel for the heat produced by the impact of the system to be expelled out of the system. The movement equations of the atoms in the thermal layer were incorporated into the Nosé–Hoover thermostat of NVT ensemble [56], while the motion equations of dynamic layer and particle were incorporated into the NVE ensemble. In the NVT ensemble, a particle-substrate system was equilibrated for 20 ps. For the second simulation setting in the NVE ensemble also for dynamic layer, an MD simulation was performed that removes the heat coupling and processes the whole system as thermally isolated. The particle spray velocity ranges from 500 m/s to 1500 m/s in the perpendicular direction to the substrate surface to preserve adhesion and prevent erosive wear behaviour [57–59]. The molecular dynamics simulation time for CGDS is 20 ps. Note, the CGDS process will not last for 20 ps, this simulation time was only allowed to ensure that the effect is completely applied is the CGDS time set. The model sizes and parameters of the simulation are presented in Table 1. The particle/substrate atomic interactions are represented by the Zhou et al. [60] embedded-atom method (EAM) potential as shown in Equation (1).

$$PE_n = \gamma_\alpha \left(\sum_{m \neq n} \rho_\beta(R_{mn}) \right) + 0.5 \sum_{m \neq n} \vartheta_{\alpha\beta}(R_{mn}) \tag{1}$$

where PE_n of atom m is the potential energy, R_{mn} is a distance from atoms n to m, the pair-wise potential function is denoted by $\vartheta_{\alpha\beta}$, ρ_β is the influence of atom type β to the electron-charge density at atomic n, and γ is the embedding function that denotes the energy needed to position type α of atom m in the electron cloud. For the analysis of the atomic stress, the stress tensor's six components are calculated based on the atomic viral stress spatial and temporal averages, as shown in Equation (2).

$$\sigma_{ab} = -\left(m v_a v_b + 0.5 \sum_{n=1}^{n_p} (R_{1a} F_{1b} + R_{2a} F_{2b}) \right) \tag{2}$$

where σ_{ab} is the components of the atomic stress arranged in $a,\ b(x, y\ or\ z)$, $m v_a v_b$ is the kinetic energy input and $0.5 \sum_{n=1}^{n_p} (R_{1a} F_{1b} + R_{2a} F_{2b})$ is the pair-wise energy input that is connected with the nearby atoms from $n = 1$ to n_p. The von Mises stress is mostly used to research plastic deformation in the CGDS process, as deduced in Equation (3).

$$\sigma_{va}(i) = \left(0.5 \sum (\sigma_{aa}(i) - \sigma_{bb}(i))^2 + 6 \sum \sigma_{ab}^2(i)\right)^{0.5} \tag{3}$$

where $\sigma_{va}(i)$ is the atom von Mises stress i, $\sigma_{ab}(i))$ is the component of atomic stress tensor arrange in a, b (x, y, or z). The atomicity is computed from each MD timestep using the interatomic interaction, atomic velocity and atomic distance; Equation (3) is an abridged representation, the measurement method information is in [61,62]. The common neighbor analysis (CNA) [63,64] is implemented to classify atoms into various local lattice frameworks for atomization (fcc- face-centered cubic, bcc- body-centered cubic, hcp- hexagonal close-packed, etc.), and jetting zone atoms (without lattice structures). Additionally, the dislocation extraction algorithm (DXA) [65] is used to classify the dislocation arrangement in crystals and generate dislocation segments.

Table 1. Schematic molecular dynamics simulation calculation plans.

Properties	Particle	Substrate
Material	Cu (1,908,733 atoms) Al (1,352,275 atoms) Ni (2,051,820 atoms) Ag (1,316,685 atoms)	Cu (19,873,953 atoms)
Dimensions	400 Å Diameter	700 Å × 700 Å × 600 Å
Initial impact velocity	500, 700, 1000, 1500 m/s	-
Temperature	273 K	300 K
Crystal orientation	[1 0 0], [0 1 0], [0 0 1] in x-, y-, and z-directions	
Force field	EAM/alloy	
Time step	0.001 ps (1.0 fs)	
Equilibration time	20 ps	
Dynamics time	20 ps	
Boundary condition	p p p	
Initial stand-off distance	20 Å	

3. Results and Discussion

3.1. Atomic Structure Evolution and Material Jet Initiation

Continuum models [34] recently proposed that jetting initiation in the course of the impact of single-particle is due to the pressure waves propagating and interacting with the particle/substrate interfacial region. The formation of a jet is due to the ejection of particulate material at the edge of the bonded interfacial region which is attributed to the creation of a tensile area resulting in a process of "spall" because of the tensile pressures produced by the particle/substrate edge. The interaction of the pressure wave at the particle/substrate periphery is suggested as a significant factor for jetting initiation instead of the shear localization process, and adiabatic shear instability has been suggested as a consequence, instead of the cause, of jetting. This phenomenon, powered by hydrodynamic pressures, is similar to that seen in fluid–particle impacts (Kelvin–Helmholtz instability); jetting is experiential when the velocity of the shock wave surpasses the velocity at the particle/substrate peripheries [66,67].

Therefore, the molecular dynamics simulations are conducted to analyze the function of the components of pressure wave generation and the creation of thermal boost-up region leading to adiabatic shear instability at the interface. The temperature evolution in the interfacial zone is contrasted with the impacts of various ductile materials such as Al/Cu, Ni/Cu, Cu/Cu, and Ag/Cu at 1000 m/s. The impact of an Ag particle creates the highest pressure wave propagating into the substrate and particle from the point of impact till around 15 ps, resulting in deformation of particles and substrate as well as heat generation followed by that of Cu, Ni, and Al impact, respectively, as shown in Figure 2. Comparative snapshots of jet-forming microstructure are shown in Figure 3 for the material combination.

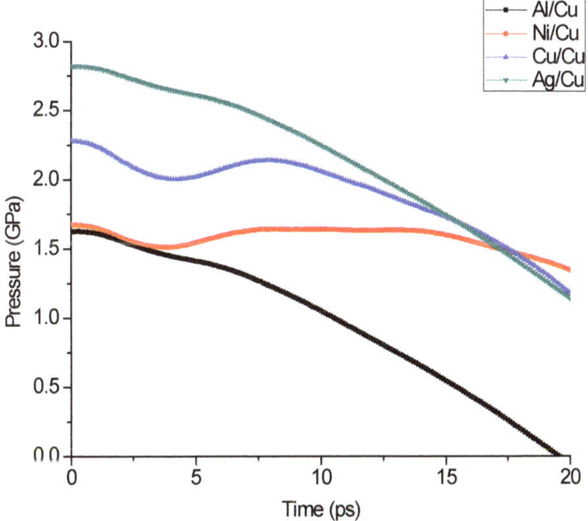

Figure 2. The pressure wave evolution at the periphery of the particle/substrate interfacial region at 1000 m/s of single-particle impact.

In Figure 4, the evolution of pressure wave interaction at particle/substrate interfacial region at 1000 m/s and 273 K in the thin cross-section through the middle of the particle is shown at 10 ps. The delineation levels are selected to provide a good visual image of the interactions between the compressive shockwave and the particle/substrate periphery if any, and its position in jet initiation. The effect produces a compressive wave of approximately 2.8 GPa for the Ag particle impact at the impact speed of 1000 m/s in the impacted interface. This compressive wave moves via the particle at the rear as well as through the interface of particle/substrate region, in conflict with particle/substrate lateral shear motion, as Figure 4 indicates. The shock wave reaches the periphery of the particle/substrate edge, in this case, leading to the external materials flow and causing the particle and substrate to form a jet. This jet initiation phenomenon resulted in the substrate and particle pressure drop and discharge at the interfacial zone as Figure 2 shows. In this context, previous research at the continuum level typically characterizes the mechanism of jetting associate with a particle only, while the MD simulations also display the position of the substrate in jet formation. In comparison, the Al/Cu, Ni/Cu, Cu/Cu, and Ag/Cu impact produces a maximum pressure wave of 1.63 GPa, 1.68 GPa, 2.28 GPa, and 2.82 GPa, respectively. The pressure wave evolution in Al/Cu impact does not seem to interfere with the interface boundary of the particle/substrate periphery and a jet initiation for Al/Cu impact is not observed, as displayed in Figure 4a.

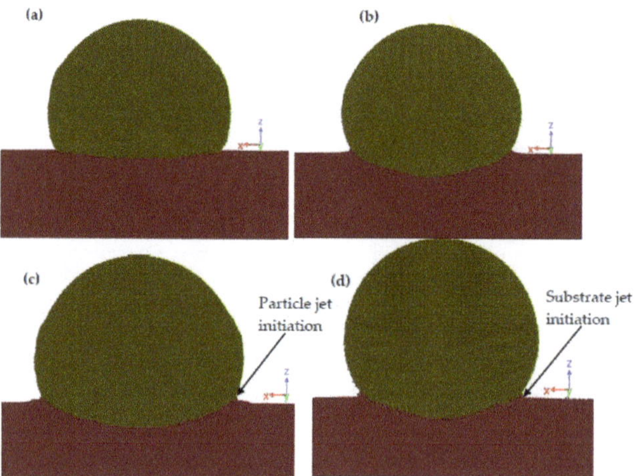

Figure 3. Splat morphologies of different material combination at 20 ps: (**a**) Al/Cu (showing no jet) for the whole simulation period (**b**) Cu/Cu, (**c**) Ni/Cu, and (**d**) Ag/Cu (b-c showing jet initiation) for 273 K.

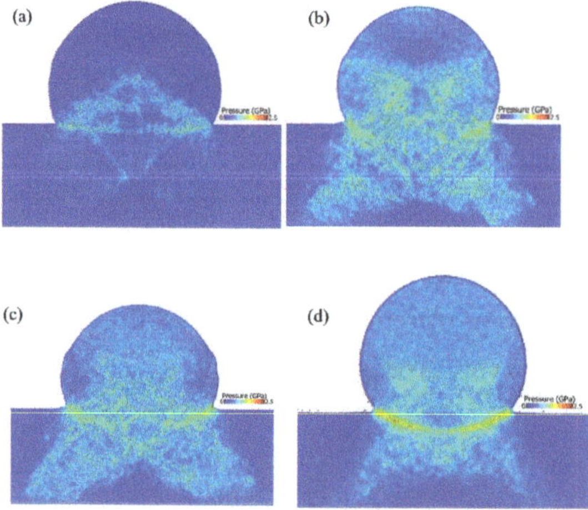

Figure 4. The evolution of pressure wave interaction at particle/substrate interfacial region at 1000 m/s and 273 K for (**a**) Al/Cu (**b**) Ni/Cu and (**c**) Cu/Cu, and (**d**) Ag/Cu (after impact at 10 ps).

While the position of the pressure wave interactions at jet initiation build-up is important, the full picture of this phenomenon should be given to further variable trends such as temperatures and von Mises stresses (flow stress) evolution. The progressive temperatures evolution at 1000 m/s impact velocity of single-particle effect for different material combinations are illustrated in Figure 5 for 20 ps simulation time. The Ag/Cu impact produces the highest interface temperatures in the particle/substrate impacting region. The temperatures in these regions are about 460 K, which is lower than the melting temperature of Ag estimated by the force field (interatomic potential). The existence and position of these elevated temperatures zones make the flow of material at the periphery of the particle/substrate interface easier. Figure 5 shows that while the high-temperature values may occur, leading to substantial material softening, the temperature gradually decreasing after reaching the peak while the outward material flow continues. The temperature spike is instigated by the emergence of the plastic deformation wave at the periphery of the particle/substrate interface, and the emergence of the waves causes the material to flow outward and the temperature drops at the interfacial edge. The corresponding flow stress evolution (von Mises) is also displayed in Figure 6 with respect to time. For clarity of visualization, atoms with the non-fcc structure that are defined in the adaptive common neighbor algorithm (CAN) are shown. At the peripheries of the contact zone of particle/substrate interface, the maximum value of von Mises stress value for the particle impact is approximately 28.2 GPa, which can be related to imminent yield and propagated rapidly within the particle with progressive flattening. In the regions near the boundary, material flow stress also increases, radially reduces to the middle of the particle interface, and falls after reaching a peak, as shown in Figure 6.

The Radial Distribution Function (RDF) is also available for testing the atomic crystallinity and providing an average, global insight of the structure of atoms in the required region of importance. The measured radial distribution function for the jetting region strongly demonstrates an amorphous structure of various combinations of materials as shown in Figure 7 after 20 ps in all particles. The RDFs results in the particle jet zone are nearly identical as shown in Figure 7. The first peak was observed at 2.50 Å, which is the closest neighbor range between the impacted pair atoms. After the initial peak comes a weak peak at approximately 3.5 Å, which is nearly close to both the gaps between pairs of atoms. Then, at 4.40 Å, 5.32 Å, and 5.75 Å, are the next three separate peaks, suggesting the development of an amorphous structure, in agreement with earlier experimental findings [68–70].

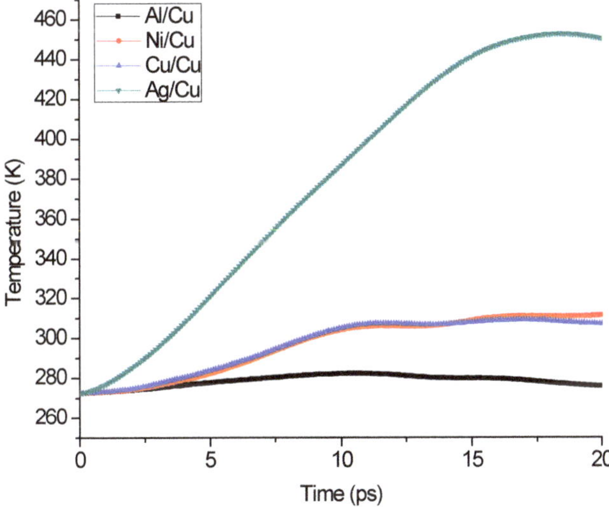

Figure 5. The temporal temperature evolution at the periphery of the particle/substrate interfacial region at 1000 m/s of single-particle impact.

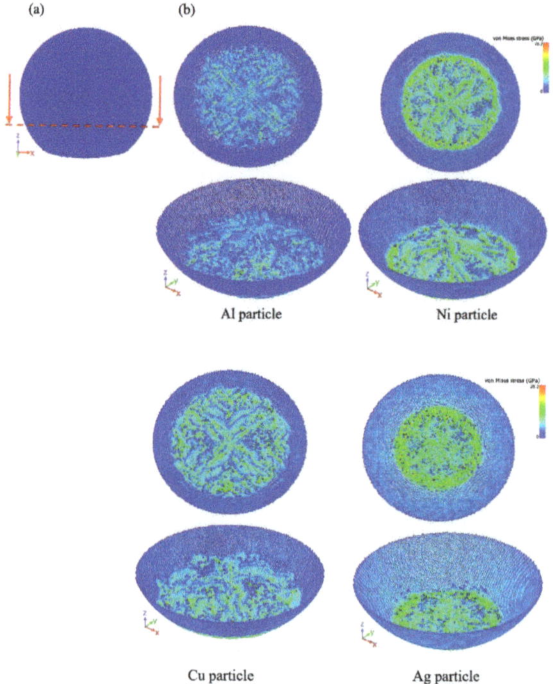

Figure 6. (**a**) Deformed particle diagram where arrows show the observation path and (**b**) von Mises stress evolution on the particle bottom after 5 ps in CGDS of the defects induced by impact.

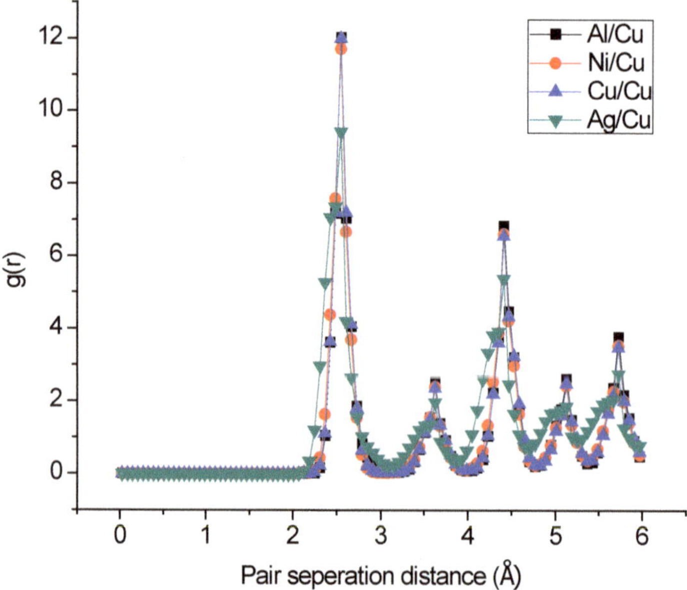

Figure 7. Radial distribution function with a different material combination of Al/Cu, Ni/Cu, Cu/Cu, and Ag/Cu after the cold gas dynamic spray at 20 ps.

3.2. Material Dislocation Plasticity

As is evident in Figure 8, plasticity of material dislocation at 10 ps on the circular side of its flattened base for which heterogeneous nucleation of Shockley with Burgers vector 1/6 ⟨1 1 2⟩ leading partial dislocation segments was observed. In the particle/substrate interfacial zone, the evolution of the microstructure corresponds to various ductile materials, where the perfect dislocation, the stair dislocations, and partial dislocations in Shockley is colored in blue, purple, and green, respectively. The yellow and sky-blue dislocation segment corresponds to the Hirth and Frank type, respectively. In the first step, Shockley partial dislocations emerge successively from the particle/substrate contact surface and disperse within the substrate and the particle. Then, with the growth of plastic deformation, the Shockley partial dislocations at 10 ps cover the particle/substrate interfacial zone. These imperfect dislocations normally act as carriers of energy for the face-centered cubic (fcc) system's intrinsic stacking faults [71]. With cold gas dynamic spray proceeding, the Burgers vector of 1/6 ⟨1 1 2⟩ with Shockley partial dislocations intersect and change to be Burgers vector of 1/6 ⟨1 1 0⟩ with the Stair-rod dislocations. Within the particle/substrate contact zone, the Burgers vector of 1/2 ⟨1 1 0⟩ appears, which is the perfect dislocation. The atoms on the circular edge of the particle flattened base steadily grow with the unstructured form of the atomic crystal. Moreover, dislocations are likely to move to the middle of the Al particle, to the edge and bottom of the Ni and Cu particle, as well as to the bottom of the Ag particle, as seen in Figure 8a–d.

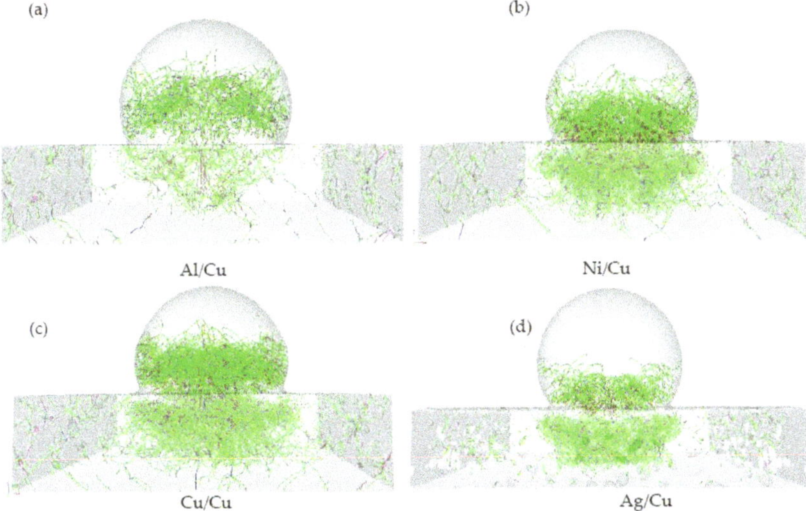

Figure 8. The evolution of dislocation segment in the particle/substrate interfacial zone during the cold gas dynamic spray processes at 10 ps for particles of (**a**) Al (**b**) Ni (**c**) Cu and (**d**) Ag impacting Cu substrate.

In all particles, during the particle impact at 1000 m/s, several dislocation segments emerge, which are due to severe plastic deformations as demonstrated in Figure 9. Additionally, on the substrate surface, the lower atoms at the base of the particles form a metal-to-metal coating region. Owing to the inadequate slip on the middle part of the Al particle, there is a significant surface protrusion in the final Al particle designated at the x–z plane, which forms a "peak-shaped" coating configuration and the ultimate particle shape is rectangular in the x–z region. For the Ni particle, the absolute form of the particle on the x–z plane is comparatively flat because of the adequate slip along the direction of material flow, and the final structure in x–z plane is like a mushroom. For the Cu and Ag particle, the final structure in the x–z plane is pyramidal and hemispheric.

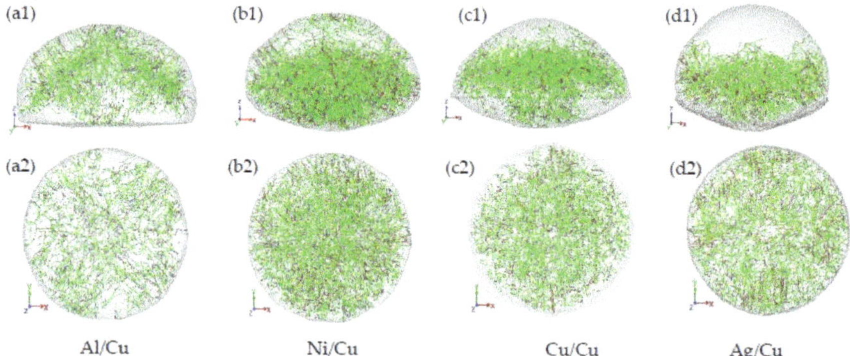

Figure 9. Cross-section view (**a1–d1**) and plan view (**a2–d2**) for the evolution of dislocation segment in the impacting particles of (**a1,a2**) Al/Cu, (**b1,b2**) Ni/Cu, (**c1,c2**) Cu/Cu, and (**d1,d2**) Ag/Cu after cold gas dynamic at 15 ps and 1000 m/s.

Figure 10 demonstrates the time-dependence of the dislocation density for different particles at 1000 m/s impact velocity. The Ni particle plastic deformation rate is the maximum following the dislocation segment evolution as shown in Figure 10a, followed by the Cu particle, then the Ag particle, and the Al particle is the lowest. In Figure 10c, the evolution of the particle dislocation density during cold gas dynamic spraying is shown between 0 ps and 20 ps. In all the different material combinations, the displacement density of all particles is near zero at the early stage of cold gas dynamic spray between 0 ps and 2 ps because no major plastic deformation exists. Then the dislocation density within various material particles from 2–20 ps increases steadily at different rates until it reaches the equilibrium after impact. The dislocation growth rate in Ni particles is about 55.4%, 33.8%, and 10.7% higher than in Al, Ag and Cu particles, respectively. In Al particle atoms, the kinetic energy is used mainly to transfer atoms along with the position of longitudinal intrinsic stacking faults. However, in Ag particles, a portion of the particle atoms' kinetic energy is utilized in the particle/substrate atomic motion, while the particle/substrate interaction, creating several dislocations by consuming the kinetic energy that remains. The particle atoms' kinetic energy of the Ni and Cu particle is primarily absorbed by the intense interaction between the intrinsic stacking faults in the lower part of the particle, which increases the dislocation density rapidly. Figure 10b shows the dislocation segments distribution in the Al, Ni, Cu and Ag particles at 20 ps. The dislocation density increases to the value of 2.9×10^{16} m^{-2}, 4.3×10^{16} m^{-2}, 5.8×10^{16} m^{-2}, and 6.5×10^{16} m^{-2} for Al, Ag, Cu, and Ni particles, respectively, then decreases slightly. The extreme plastic deformation at the interface of the substrate and material particle contributes to the creation of a jet region around the interface region of particle/substrate.

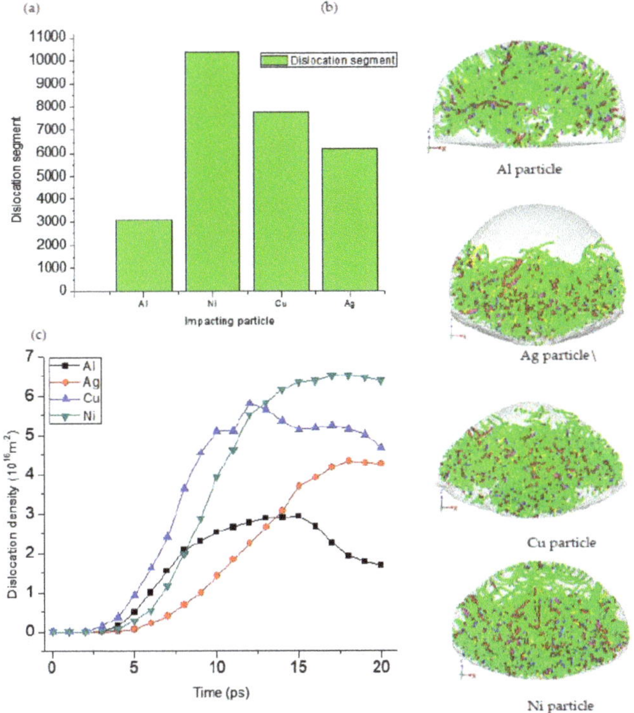

Figure 10. (**a**) Total dislocation segment of the particle at 20 ps (**b**) Dislocations segment distribution in the Al, Ag, Cu, Ni particles at 20 ps. (**c**) Particle dislocation density at 1000 m/s, 273 K and 20 ps after impact for the Al, Ni, Cu, and Ag particles.

3.3. Particle Impact Velocity Effect on Microstructure Evolution

The microstructural snapshot of the full Ni splat region coated at different times on the Cu substrate for an impact speed of 1000 m/s is shown in Figure 11a–c. The local atomic structure is colored based on common neighbor analysis (CNA) approach [63,64]: FCC (green), HCP (red), BCC (blue) and grey (atoms without crystalline structure). The circles designated the jet formation at the periphery as the atoms flow upward from the substrate. The snapshots presenting the particle and substrate separately of the illustrative atoms. The material jet initiation at the periphery is often correlated with the outward flow of atoms as shown by the circles from the substrate.

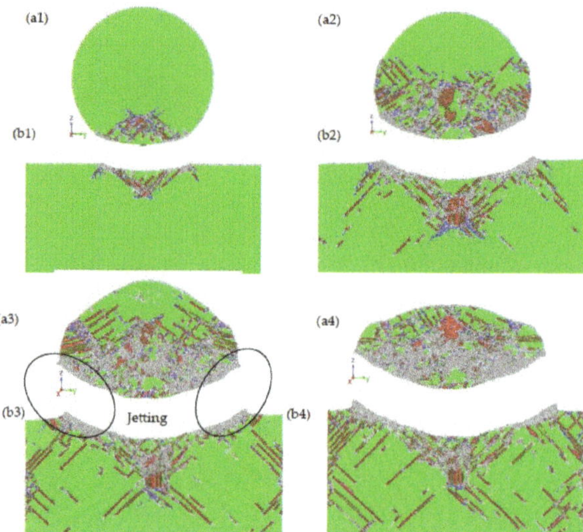

Figure 11. The microstructure snapshots of the Ni particle (**a1–a4**) and Cu substrate (**b1–b4**) at times of (**a1**,**b1**) 5 ps, (**a2**,**b2**) 10 ps, (**a3**,**b3**) 15 ps, and (**a4**,**b4**) 20 ps. (after impact) at 1000 m/s. The local atomic structure: FCC (green), HCP (red), blue (BCC), and grey (atoms without crystalline structure). The circles designated the jet formation at the periphery as the atoms flow upward from the substrate.

During MD simulations of single material impact for a particle impact velocity of 500 m/s, 700 m/s, 1000 m/s, and 1500 m/s, the microstructure evolution and recrystallization behavior are also investigated. The recrystallization potential varies at impact speeds as establish in the cold dynamic gas sprayed splats created during 20 ps, which shows the local atomic structure at 273 K in Figure 12. With a particle velocity of the impact of 500 m/s, recrystallized structures are very few, as indicated in Figure 12a, in contrast with the 1500 m/s particle impact velocity with large structural transformation as displayed in Figure 12d. The increased number of new atomic structures can be associated with the reliance of recrystallization phenomenon associated with strain energy to indicate the unique recrystallization mechanisms [69,72] involved in the molecular dynamics predicted microstructures given the extreme deformation of the atomic structure at the interface. In addition to the recrystallization in the creation of new structures, atomic structure boundary migration [73] could also play a critical position. For convenience, in this manuscript, the development of a new atomic structure is called recrystallization. Metal plastic deformation is more severe with a higher impact velocity of 1500 m/s, and therefore greater temperatures are produced during the impact time at the particle/substrate interfacial region as shown in Figure 13. Some experimental cold gas dynamic sprays have achieved a similar consensus on the recrystallization phenomenon [74–76]. In the past, the embedded atomic potential (EAM/alloy) [60] used in this study has shown a recrystallization of aluminum uniaxial loading, demonstrating its ability to capture the phenomenon accurately [77]. Figure 14 shows that, at the interface, some of this recrystallized atomic structure is split between particle and substrate. These findings show that recrystallization at some region at the interfacial zones will help form the metallurgical bond between the substrate and the particle.

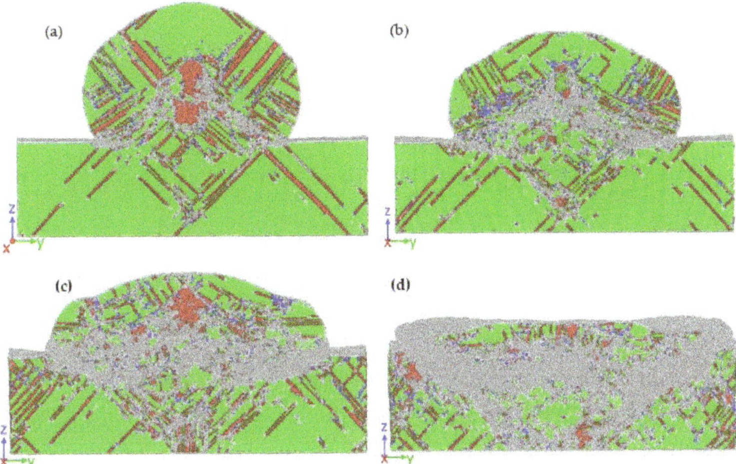

Figure 12. Microstructural snapshots presenting a structural transformation of a cold gas dynamic spray splat at 20 ps for: (**a**) 500 m/s, (**b**) 700 m/s, (**c**) 1000 m/s, and (**d**) 1500 m/s.

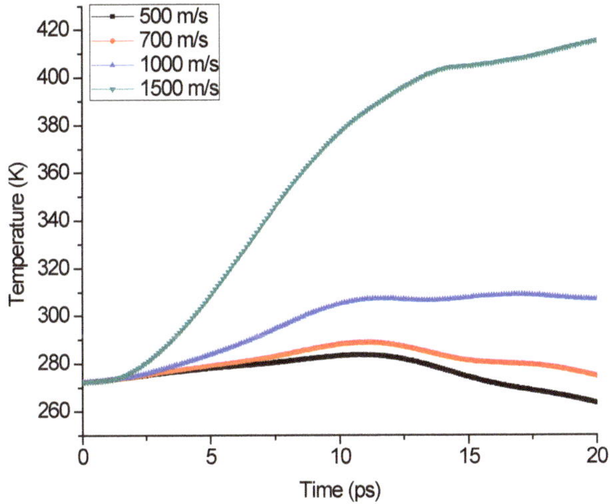

Figure 13. The temperature evolution at the periphery of the Cu/Cu interfacial region at different impacting velocity of 500 m/s, 700 m/s, 1000 m/s, and 1600 m/s of single particle impact.

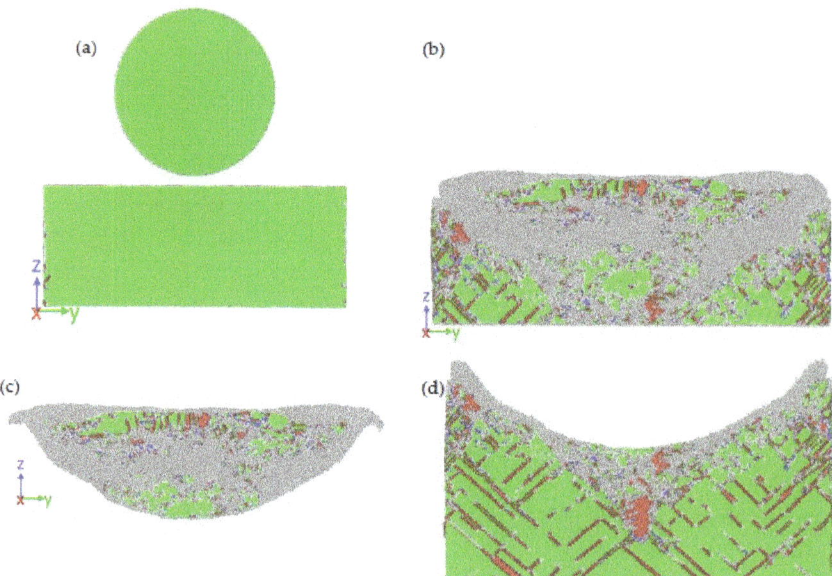

Figure 14. Microstructural snapshots presenting a structural transformation (**a**) at 0 ps (**b**) at 20 ps of a cold gas dynamic spray splat at 1500 m/s (**c**) splat microstructure at 20 ps (**d**) substrate microstructure at 20 ps. Structural transformation at the interfacial zone between the substrate and particle after impact is evident here.

4. Conclusions

Cold gas dynamic spray impacts on pure copper substrates of nanoscale ductile materials such as Al, Ni, Cu, and Ag are modelled using a molecular dynamic simulation (MD) system for differing material combinations, particle impact velocities, and microstructures. The post-impact behaviour is objectively studied by monitoring the distribution of the thermomechanical variable such as pressure wave, temperature, stress, dislocation density, and microstructural features such as local atomic structural transformation at the nanoscale. We studied the position and the propagation of the compressive shock wave in activating a jet at the interface of particle/substrate impact region.

The findings of the simulation show that the jetting phenomenon, as the material deforms during impact, is due to the interaction between the velocity of the shock wave at the interface of the particle/substrate interfacial zone and the velocity at the edge of particle/substrate interface. The effect produces a compressive wave of approximately 2.8 GPa for the Ag particle impact at the impact speed of 1000 m/s in the impacted zone. This compressive wave moves via the particle at the rear as well as through the interface of particle/substrate region, in conflict with particle/substrate lateral shear motion. The shock wave reaches the periphery of the particle/substrate edge, in this case, leading to the external materials flow and causing the particle and substrate to form a jet. The pressure wave evolution in Al/Cu impact does not seem to interfere with the interface boundary of the particle/substrate periphery and a jet initiation for Al/Cu impact is not observed. The impact of an Ag particle creates the highest-pressure wave propagating into the substrate and particle from the point of impact till around 15 ps, resulting in deformation of particles and substrate as well as heat generation followed by that of Cu, Ni, and Al impact, respectively.

The deformation behavior for individual impacting particles and subsequent dislocation evolution allowed us to classify the stages of deformation into three, after the impact analysis of particle/substrate impact. Firstly, plastic deformation began by the slide of dislocations and nucleation at the exterior end of the particle that is in close contact with the surface of the substrates at impact. Thereafter, with additional deformation, the analysis of the dense dislocation segment when comparing the initial and the final microstructures of the splat show dislocation network formation formed at the exterior bottom of the particles, and the upper region remains essentially undeformed. Finally, as seen in the finite element method and microscopy experiments, the upper part the particle deformed as well, causing the usual flattened splat shape. The dislocation density increases to the value of 2.9×10^{16} m^{-2}, 4.3×10^{16} m^{-2}, 5.8×10^{16} m^{-2}, and 6.5×10^{16} m^{-2} for Al, Ag, Cu, and Ni particles, respectively. The extreme plastic deformation at the interface of the substrate and material particle contributes to the creation of a jet region around the interface region of particle/substrate. The dislocation growth rate in Ni particles is about 55.4%, 33.8%, and 10.7% higher than in Al, Ag, and Cu particles, respectively.

A contrast of the original and final microstructures of the splat shows the creation, near the interface, of "new" atomic structure, which is certified by a combination of atomic boundary mobility stress-led and recrystallization of particle/substrate mechanisms. The recrystallization potential varies with the impact velocities, the thermal evolution and the stored strain energy as seen in the CGDS splats created during 20 ps, which show the local atomic structure. Some of these new structures are established on the bond line, which facilitates close interaction and strengthens CGDS splat bond. These findings expand our awareness about the processes to reinforce cold gas dynamic sprayed deposits from a modelling perspective. CGDS also qualifies as an additive manufacturing procedure to achieve the optimal strength of the deposit (coating) by the interface nucleation of finer atomic structures and the subsequent atomic boundary strengthening.

Author Contributions: S.T.O.: Conceptualization, Methodology, Investigation, Software, Visualization, Writing-Original draft. T.-C.J.: Resources, Reviewing and Editing and Supervision. All authors have read and agreed to the published version of the manuscript.

Funding: This research received no external funding.

Acknowledgments: The authors would like to acknowledge the financial support from the University Research Committee of the University of Johannesburg and the National Research Foundation of South Africa.

Conflicts of Interest: The authors declare no conflict of interest.

References

1. Ganesan, A.; Affi, J.; Yamada, M.; Fukumoto, M. Bonding behavior studies of cold sprayed copper coating on the PVC polymer substrate. *Surf. Coat. Technol.* **2012**, *207*, 262–269. [CrossRef]
2. Khodabakhshi, F.; Marzbanrad, B.; Jahed, H.; Gerlich, A. Interfacial bonding mechanisms between aluminum and titanium during cold gas spraying followed by friction-stir modification. *Appl. Surf. Sci.* **2018**, *462*, 739–752. [CrossRef]
3. Lu, J.; Zhang, H.; Chen, Y.; Zhao, X.; Guo, F.; Xiao, P. Effect of microstructure of a NiCoCrAlY coating fabricated by high-velocity air fuel on the isothermal oxidation. *Corros. Sci.* **2019**, *159*. [CrossRef]
4. Raoelison, R.; Xie, Y.; Sapanathan, T.; Planche, M.; Kromer, R.; Costil, S.; Langlade, C. Cold gas dynamic spray technology: A comprehensive review of processing conditions for various technological developments till to date. *Addit. Manuf.* **2018**, *19*, 134–159. [CrossRef]
5. Bagherifard, S.; Monti, S.; Zuccoli, M.V.; Riccio, M.; Kondás, J.; Guagliano, M. Cold spray deposition for additive manufacturing of freeform structural components compared to selective laser melting. *Mater. Sci. Eng. A* **2018**, *721*, 339–350. [CrossRef]
6. Oyinbo, S.T.; Jen, T.-C. Investigation of the process parameters and restitution coefficient of ductile materials during cold gas dynamic spray (CGDS) using finite element analysis. *Addit. Manuf.* **2020**, *31*, 100986. [CrossRef]
7. Oyinbo, S.; Jen, T.-C. A comparative review on cold gas dynamic spraying processes and technologies. *Manuf. Rev.* **2019**, *6*, 25. [CrossRef]
8. Irissou, E.; Legoux, J.-G.; Ryabinin, A.N.; Jodoin, B.; Moreau, C. Review on Cold Spray Process and Technology: Part I—Intellectual Property. *J. Therm. Spray Technol.* **2008**, *17*, 495–516. [CrossRef]
9. Van Steenkiste, T.; Smith, J.; Teets, R. Aluminum coatings via kinetic spray with relatively large powder particles. *Surf. Coat. Technol.* **2002**, *154*, 237–252. [CrossRef]

10. Assadi, F.H.; Gartner, T.; Stoltenhoff, H.K. Bonding mechanism in cold gas spraying. *Acta Mater.* **2003**, *51*, 4379–4394. [CrossRef]
11. Grujicic, M.; Zhao, C.; Tong, C.; Derosset, W.; Helfritch, D. Analysis of the impact velocity of powder particles in the cold-gas dynamic-spray process. *Mater. Sci. Eng. A* **2004**, *368*, 222–230. [CrossRef]
12. Nasim, M.; Vo, T.; Mustafi, L.; Kim, B.; Lee, C.S.; Chu, W.-S.; Chun, D.-M. Deposition mechanism of graphene flakes directly from graphite particles in the kinetic spray process studied using molecular dynamics simulation. *Comput. Mater. Sci.* **2019**, *169*. [CrossRef]
13. Rafaja, D.; Schucknecht, T.; Klemm, V.; Paul, A.; Berek, H. Microstructural characterisation of titanium coatings deposited using cold gas spraying on Al_2O_3 substrates. *Surf. Coat. Technol.* **2009**, *203*, 3206–3213. [CrossRef]
14. Grujicic, M.; Saylor, J.; Beasley, D.; Derosset, W.; Helfritch, D. Computational analysis of the interfacial bonding between feed-powder particles and the substrate in the cold-gas dynamic-spray process. *Appl. Surf. Sci.* **2003**, *219*, 211–227. [CrossRef]
15. Zhang, D.; Shipway, P.; McCartney, D. Cold Gas Dynamic Spraying of Aluminum: The Role of Substrate Characteristics in Deposit Formation. *J. Therm. Spray Technol.* **2005**, *14*, 109–116. [CrossRef]
16. Kim, D.-Y.; Park, J.-J.; Lee, J.-G.; Kim, D.; Tark, S.J.; Ahn, S.; Yun, J.H.; Gwak, J.; Yoon, K.H.; Chandra, S.; et al. Cold Spray Deposition of Copper Electrodes on Silicon and Glass Substrates. *J. Therm. Spray Technol.* **2013**, *22*, 1092–1102. [CrossRef]
17. King, P.C.; Zahiri, S.; Jahedi, M.; Friend, J. Aluminium coating of lead zirconate titanate—A study of cold spray variables. *Surf. Coat. Technol.* **2010**, *205*, 2016–2022. [CrossRef]
18. Hussain, T.; McCartney, D.; Shipway, P. Bonding between aluminium and copper in cold spraying: Story of asymmetry. *Mater. Sci. Technol.* **2012**, *28*, 1371–1378. [CrossRef]
19. Oyinbo, S.; Jen, T.-C. Feasibility of numerical simulation methods on the Cold Gas Dynamic Spray (CGDS) Deposition process for ductile materials. *Manuf. Rev.* **2020**, *7*, 24. [CrossRef]
20. Seo, D.; Sayar, M.; Ogawa, K. SiO_2 and $MoSi_2$ formation on Inconel 625 surface via SiC coating deposited by cold spray. *Surf. Coat. Technol.* **2012**, *206*, 2851–2858. [CrossRef]
21. Burlacov, I.; Jirkovský, J.; Kavan, L.; Ballhorn, R.; Heimann, R.B. Cold gas dynamic spraying (CGDS) of TiO_2 (anatase) powders onto poly(sulfone) substrates: Microstructural characterisation and photocatalytic efficiency. *J. Photochem. Photobiol. A Chem.* **2007**, *187*, 285–292. [CrossRef]
22. Amirthan, G.; Yamada, M.; Fukumoto, M. Cold Spray Coating Deposition Mechanism on the Thermoplastic and Thermosetting Polymer Substrates. *J. Therm. Spray Technol.* **2013**, *22*, 1275–1282. [CrossRef]
23. Lupoi, R.; O'Neill, W. Deposition of metallic coatings on polymer surfaces using cold spray. *Surf. Coat. Technol.* **2010**, *205*, 2167–2173. [CrossRef]
24. Zhou, X.; Chen, A.; Liu, J.; Wu, X.; Li, Y. Preparation of metallic coatings on polymer matrix composites by cold spray. *Surf. Coat. Technol.* **2011**, *206*, 132–136. [CrossRef]
25. King, P.C.; Poole, A.J.; Horne, S.; De Nys, R.; Gulizia, S.; Jahedi, M.Z. Embedment of copper particles into polymers by cold spray. *Surf. Coat. Technol.* **2013**, *216*, 60–67. [CrossRef]
26. Alhulaifi, A.S.; Buck, G.A.; Arbegast, W.J. Numerical and Experimental Investigation of Cold Spray Gas Dynamic Effects for Polymer Coating. *J. Therm. Spray Technol.* **2012**, *21*, 852–862. [CrossRef]
27. Robitaille, F.; Yandouzi, M.; Hind, S.; Jodoin, B. Metallic coating of aerospace carbon/epoxy composites by the pulsed gas dynamic spraying process. *Surf. Coat. Technol.* **2009**, *203*, 2954–2960. [CrossRef]
28. Champagne, V.K.; Helfritch, D.; Leyman, P.; Grendahl, S.; Klotz, B. Interface Material Mixing Formed by the Deposition of Copper on Aluminum by Means of the Cold Spray Process. *J. Therm. Spray Technol.* **2005**, *14*, 330–334. [CrossRef]
29. Ko, K.; Choi, J.; Lee, H. Intermixing and interfacial morphology of cold-sprayed Al coatings on steel. *Mater. Lett.* **2014**, *136*, 45–47. [CrossRef]
30. Yin, S.; Cavaliere, P.; Aldwell, B.; Jenkins, R.; Liao, H.; Li, W.; Lupoi, R. Cold spray additive manufacturing and repair: Fundamentals and applications. *Addit. Manuf.* **2018**, *21*, 628–650. [CrossRef]
31. Moridi, A.; Gangaraj, S.; Guagliano, M.; Dao, M. Cold spray coating: Review of material systems and future perspectives. *Surf. Eng.* **2014**, *30*, 369–395. [CrossRef]
32. Xie, Y.; Yin, S.; Chen, C.; Planche, M.-P.; Liao, H.; Lupoi, R. New insights into the coating/substrate interfacial bonding mechanism in cold spray. *Scr. Mater.* **2016**, *125*, 1–4. [CrossRef]

33. Grujicic, M.; Zhao, C.; Derosset, W.; Helfritch, D. Adiabatic shear instability based mechanism for particles/substrate bonding in the cold-gas dynamic-spray process. *Mater. Des.* **2004**, *25*, 681–688. [CrossRef]
34. Hassani-Gangaraj, M.; Veysset, D.; Champagne, V.K.; Nelson, K.A.; Schuh, C.A. Adiabatic shear instability is not necessary for adhesion in cold spray. *Acta Mater.* **2018**, *158*, 430–439. [CrossRef]
35. Murr, L.E.; Esquivel, E.V. Observations of common microstructural issues associated with dynamic deformation phenomena: Twins, microbands, grain size effects, shear bands, and dynamic recrystallization. *J. Mater. Sci.* **2004**, *39*, 1153–1168. [CrossRef]
36. Rahmati, S.; Zúñiga, A.; Jodoin, B.; Veiga, R. Deformation of copper particles upon impact: A molecular dynamics study of cold spray. *Comput. Mater. Sci.* **2020**, *171*, 109219. [CrossRef]
37. Oyinbo, S.; Jen, T.-C.; Aasa, S.A.; Abegunde, O.; Zhu, Y. Development of palladium nanoparticles deposition on a copper substrate using a molecular dynamic (MD) simulation: A cold gas dynamic spray process. *Manuf. Rev.* **2020**, *7*, 29. [CrossRef]
38. Bae, G.; Kumar, S.; Yoon, S.; Kang, K.; Na, H.; Kim, H.-J.; Lee, C. Bonding features and associated mechanisms in kinetic sprayed titanium coatings. *Acta Mater.* **2009**, *57*, 5654–5666. [CrossRef]
39. Oyinbo, S.T.; Jen, T.-C.; Zhu, Y.; Ajiboye, J.S.; Ismail, S.O. Atomistic simulations of interfacial deformation and bonding mechanism of Pd-Cu composite metal membrane using cold gas dynamic spray process. *Vacuum* **2020**, *182*, 109779. [CrossRef]
40. Wang, Q.; Birbilis, N.; Zhang, M.-X. Interfacial structure between particles in an aluminum deposit produced by cold spray. *Mater. Lett.* **2011**, *65*, 1576–1578. [CrossRef]
41. Guetta, S.; Berger, M.-H.; Borit, F.; Guipont, V.; Jeandin, M.; Boustie, M.; Ichikawa, Y.; Sakaguchi, K.; Ogawa, K. Influence of Particle Velocity on Adhesion of Cold-Sprayed Splats. *J. Therm. Spray Technol.* **2009**, *18*, 331–342. [CrossRef]
42. Henao, J.; Concustell, A.; Dosta, S.; Bolelli, G.; Cano, I.; Lusvarghi, L.; Guilemany, J. Deposition mechanisms of metallic glass particles by Cold Gas Spraying. *Acta Mater.* **2017**, *125*, 327–339. [CrossRef]
43. Oyinbo, S.T.; Jen, T.-C. Molecular dynamics investigation of temperature effect and surface configurations on multiple impacts plastic deformation in a palladium-copper composite metal membrane (CMM): A cold gas dynamic spray (CGDS) process. *Comput. Mater. Sci.* **2020**, *185*, 109968. [CrossRef]
44. Fukumoto, M.; Mashiko, M.; Yamada, M.; Yamaguchi, E. Deposition Behavior of Copper Fine Particles onto Flat Substrate Surface in Cold Spraying. *J. Therm. Spray Technol.* **2009**, *19*, 89–94. [CrossRef]
45. Zhao, P.; Zhang, Q.; Guo, Y.; Liu, H.; Deng, Z. Atomic simulation of crystal orientation effect on coating surface generation mechanisms in cold spray. *Comput. Mater. Sci.* **2020**, *184*, 109859. [CrossRef]
46. Salehinia, I.; Lawrence, S.K.; Bahr, D. The effect of crystal orientation on the stochastic behavior of dislocation nucleation and multiplication during nanoindentation. *Acta Mater.* **2013**, *61*, 1421–1431. [CrossRef]
47. Basu, S.; Wang, Z.; Saldana, C. Anomalous evolution of microstructure and crystallographic texture during indentation. *Acta Mater.* **2016**, *105*, 25–34. [CrossRef]
48. Oyinbo, S.; Jen, T.-C. A Molecular Dynamics Investigation of the Temperature Effect on the Mechanical Properties of Selected Thin Films for Hydrogen Separation. *Membranes* **2020**, *10*, 241. [CrossRef]
49. Hassani-Gangaraj, M.; Veysset, D.; Nelson, K.A.; Schuh, C.A. In-situ observations of single micro-particle impact bonding. *Scr. Mater.* **2018**, *145*, 9–13. [CrossRef]
50. King, P.C.; Zahiri, S.H.; Jahedi, M. Microstructural Refinement within a Cold-Sprayed Copper Particle. *Met. Mater. Trans. A* **2009**, *40*, 2115–2123. [CrossRef]
51. Irissou, E.; Legoux, J.-G.; Arsenault, B.; Moreau, C. Investigation of Al-Al_2O_3 Cold Spray Coating Formation and Properties. *J. Therm. Spray Technol.* **2007**, *16*, 661–668. [CrossRef]
52. King, P.C.; Busch, C.; Kittel-Sherri, T.; Jahedi, M.; Gulizia, S. Interface melding in cold spray titanium particle impact. *Surf. Coat. Technol.* **2014**, *239*, 191–199. [CrossRef]
53. Schöner, C.; Pöschel, T. Orientation-dependent properties of nanoparticle impact. *Phys. Rev. E* **2018**, *98*, 022902. [CrossRef] [PubMed]
54. Plimpton, S. Fast Parallel Algorithms for Short-Range Molecular Dynamics. *J. Comput. Phys.* **1995**, *117*, 1–19. [CrossRef]
55. Stukowski, A. Visualization and analysis of atomistic simulation data with OVITO–the Open Visualization Tool. *Model. Simul. Mater. Sci. Eng.* **2009**, *18*. [CrossRef]
56. Braga, C.; Travis, K.P. A configurational temperature Nosé-Hoover thermostat. *J. Chem. Phys.* **2005**, *123*, 134101. [CrossRef]

57. Hassani-Gangaraj, M.; Veysset, D.; Champagne, V.K.; Nelson, K.A.; Schuh, C.A. Response to Comment on Adiabatic shear instability is not necessary for adhesion in cold spray. *Scr. Mater.* **2019**, *162*, 515–519. [CrossRef]
58. Li, W.-Y.; Gao, W. Some aspects on 3D numerical modeling of high velocity impact of particles in cold spraying by explicit finite element analysis. *Appl. Surf. Sci.* **2009**, *255*, 7878–7892. [CrossRef]
59. Ghelichi, R.; Bagherifard, S.; Guagliano, M.; Verani, M. Numerical simulation of cold spray coating. *Surf. Coat. Technol.* **2011**, *205*, 5294–5301. [CrossRef]
60. Zhou, X.W.; Johnson, R.A.; Wadley, H.N.G. Misfit-energy-increasing dislocations in vapor-deposited CoFe/NiFe multilayers. *Phys. Rev. B* **2004**, *69*, 1–10. [CrossRef]
61. Guo, Y.-B.; Xu, T.; Li, M. Generalized type III internal stress from interfaces, triple junctions and other microstructural components in nanocrystalline materials. *Acta Mater.* **2013**, *61*, 4974–4983. [CrossRef]
62. Guo, Y.-B.; Xu, T.; Li, M. Hierarchical dislocation nucleation controlled by internal stress in nanocrystalline copper. *Appl. Phys. Lett.* **2013**, *102*, 241910. [CrossRef]
63. Faken, D.; Jónsson, H. Systematic analysis of local atomic structure combined with 3D computer graphics. *Comput. Mater. Sci.* **1994**, *2*, 279–286. [CrossRef]
64. Honeycutt, J.D.; Andersen, H.C. Molecular dynamics study of melting and freezing of small Lennard-Jones clusters. *J. Phys. Chem.* **1987**, *91*, 4950–4963. [CrossRef]
65. Stukowski, A.; Bulatov, V.V.; Arsenlis, A. Automated identification and indexing of dislocations in crystal interfaces. *Model. Simul. Mater. Sci. Eng.* **2012**, *20*. [CrossRef]
66. Dear, J.P.; Field, J.E. High-speed photography of surface geometry effects in liquid/solid impact. *J. Appl. Phys.* **1988**, *63*, 1015–1021. [CrossRef]
67. Field, J.E.; Dear, J.P.; Ogren, J.E. The effects of target compliance on liquid drop impact. *J. Appl. Phys.* **1989**, *65*, 533–540. [CrossRef]
68. Li, W.Y.; Yin, S.; Wang, X.-F. Numerical investigations of the effect of oblique impact on particle deformation in cold spraying by the SPH method. *Appl. Surf. Sci.* **2010**, *256*, 3725–3734. [CrossRef]
69. Rokni, M.R.; Zarei-Hanzaki, A.; Abedi, H. Microstructure evolution and mechanical properties of back extruded 7075 aluminum alloy at elevated temperatures. *Mater. Sci. Eng. A* **2012**, *532*, 593–600. [CrossRef]
70. Arabgol, Z.; Vidaller, M.V.; Assadi, H.; Gärtner, F.; Klassen, T. Influence of thermal properties and temperature of substrate on the quality of cold-sprayed deposits. *Acta Mater.* **2017**, *127*, 287–301. [CrossRef]
71. Hull, D.; Bacon, D.J. *Introduction to Dislocation*, 4th ed.; Woburn: Butterworth, MA, USA, 2001.
72. Rokni, M.R.; Widener, C.A.; Champagne, V. Microstructural stability of ultrafine grained cold sprayed 6061 aluminum alloy. *Appl. Surf. Sci.* **2014**, *290*, 482–489. [CrossRef]
73. Cheng, K.; Zhang, L.; Lu, C.; Tieu, K. Coupled grain boundary motion in aluminium: The effect of structural multiplicity. *Sci. Rep.* **2016**, *6*, 25427. [CrossRef] [PubMed]
74. Zou, Y.; Goldbaum, D.; Szpunar, J.A.; Yue, S. Microstructure and nanohardness of cold-sprayed coatings: Electron backscattered diffraction and nanoindentation studies. *Scr. Mater.* **2010**, *62*, 395–398. [CrossRef]
75. Rokni, M.R.; Widener, C.; Crawford, G.; West, M. An investigation into microstructure and mechanical properties of cold sprayed 7075 Al deposition. *Mater. Sci. Eng. A* **2015**, *625*, 19–27. [CrossRef]
76. Rokni, M.; Widener, C.; Crawford, G. Microstructural evolution of 7075 Al gas atomized powder and high-pressure cold sprayed deposition. *Surf. Coat. Technol.* **2014**, *251*, 254–263. [CrossRef]
77. Li, X.; Wei, Y.; Yang, W.; Gao, H. Competing grain-boundary- and dislocation-mediated mechanisms in plastic strain recovery in nanocrystalline aluminum. *Proc. Natl. Acad. Sci. USA* **2009**, *106*, 16108–16113. [CrossRef]

Article

Microstructure Control and Friction Behavior Prediction of Laser Cladding Ni35A+TiC Composite Coatings

Xu Huang [1,2], Chang Liu [1], Hao Zhang [1], Changrong Chen [1,3,*], Guofu Lian [1], Jibin Jiang [1], Meiyan Feng [1] and Mengning Zhou [1]

1 School of Mechanical & Automotive Engineering, Fujian University of Technology, Fuzhou 350118, China; huangxu@fjut.edu.cn (X.H.); changliukxm@163.com (C.L.); zhanghao573@163.com (H.Z.); gflian@mail.ustc.edu.cn (G.L.); jibinj@fjut.edu.cn (J.J.); happy_10236@126.com (M.F.); 18760233903@163.com (M.Z.)

2 Fujian Innovation Center of Additive Manufacturing, Fuzhou 350118, China

3 Digital Fujian Industrial Manufacturing IoT Lab, Fuzhou 350118, China

* Correspondence: changrong.chen@fjut.edu.cn; Tel.: +86-0591-22863232

Received: 8 July 2020; Accepted: 7 August 2020; Published: 9 August 2020

Abstract: The premise of surface strengthening and repair of high valued components is to identify the relationship between coating formulation, structure, and properties. Based on the full factorial design, the effects of process parameters (laser power, scanning speed, gas-powder flow rate, and weight fraction of TiC) on the phase composition, microstructure, and element distribution of Ni35A/TiC cladding layer were investigated, followed by the cause identification of wear behavior. Through ANOVA, the correlation was established with good prediction accuracy (R^2 = 0.9719). The most important factors affecting the wear rate of the cladding layer were recognized as laser power and particle ratio with a p-value < 0.001. The cladding layer was mainly comprised of Ni_3Fe and $TiC_{0.957}$. The excessive laser power would enhance the process of convection-diffusion of the melt pool, increase dilution, and improve wear volume. High laser power facilitates renucleation and growth of the hard phase, especially the complete growth of secondary axis dendrite for the top region. Increased TiC significantly changes the microstructure of the hard phase into a non-direction preferable structure, which prevents stress concentration at tips and further improves the mechanical properties. The research results are a valuable support for the manipulation of microstructure and prediction of wear behavior of composite cladding layer.

Keywords: laser cladding; TiC; microstructure control; wear behavior prediction

1. Introduction

As an advanced material processing technology, laser cladding is a novel surface modification technique. Due to the unique characteristics of high energy density, rapid solidification rate, and high cooling rate, laser cladding has been widely used for obtaining better wear and corrosion-resistant coatings with minimal heat affected zone, reliable metallurgical bonding, and uniformly distributed fine microstructure [1–5].

In order to further improve the mechanical and physical and chemical properties of the coating, in recent years, the research on the performance of the composite coating has attracted much attention, and the reinforced phase ceramic powder has become a research hotspot. Among them, TiC is an attractive reinforced phase for improving coating properties for its excellent physical and mechanical properties, such as high melting point, high hardness, low density, and strong covalent bonding [6,7]. Ni-based self-fluxing alloy is also frequently applied for surface strengthening due to its good wear resistance, impact resistance, and excellent fatigue behavior at high temperatures [8–10]. Compounding the two and performing laser cladding is expected to obtain a new type of composite coating with more excellent physical and chemical properties. Some scholars have conducted related research and made some progress, but the raw material formulation of the multi-phase coating and the comprehensive mechanism of the effect of process selection on tissue performance are still in doubt. Therefore, this paper designed a full-factor experiment for Ni35A/TiC composite coating and attempted to elaborate on the influence of formula and process on the structure and the corresponding relationship between microstructure and friction performance.

At present, many scholars have studied the preparation and performance optimization of TiC-reinforced ceramic composite coatings, using different raw material forms and molding methods to optimize the hardness, tensile strength, wear resistance, corrosion resistance, and other properties of the composite coatings, and they have made some progress. Hong et al. [11] used laser metal deposition technology to prepare ultrafine TiC particle-reinforced Inconel 625 composite parts. It was found that adding an ultrafine TiC particle could obtain a better columnar dendrite structure and could significantly improve the parts' tensile properties and wear resistance. Xu et al. [12] found that compared with pure Inconel 625 coating, the microhardness and tensile strength of TiC-reinforced Inconel 625 coating were significantly improved, and the addition of TiC-reinforced Inconel 625 coating also showed good corrosion resistance. Saroj et al. [13] used the TIG (Tungsten Inert Gas Welding) cladding process to prepare TiC-reinforced Inconel 825 composite coatings with different mass percentages (20%, 40%, 60%). The results showed that TiC had an important effect on the morphology of the composite coating. The SEM structure and friction and wear analysis showed that as the TiC content in the TiC/Inconel 825 coating increased, the friction coefficient of the composite coating decreased significantly. At the same time, through organization identification and performance analysis, scholars have further clarified the evolution and strengthening mechanism of the composite phase in the cladding layer. Sahoo et al. [14] adopted a pre-coating method and obtained a wear-resistant TiC/Ni composite coating using the TIG cladding process. It was found that the presence of TiC, Ni, and some intermetallic compounds in the coating affected the excellent interface bonding and coating. The main reason for the layer to obtain better wear resistance and TiC/Ni composite coating wear resistance is 70 times higher than the AlS1304 substrate.

Although a lot of research has been conducted on the performance of TiC composite coatings, there are few reports on the multi-factor coupling relationship between the laser cladding process parameters and the composite coating wear rate. After the pyrolysis of the hard phase in the cladding process, the directional growth of the hard phase in the molten pool with a significant gradient distribution of the temperature field brings the tip stress problem to be solved urgently. Therefore, in this work, a full factorial experimental design ($L4^2$) was proposed to investigate the effects of process parameters together with the weight fraction of TiC on the wear rate and microstructure evolution of composite coating.

2. Materials and Methods

The substrates were made of AISI 1045 steel with dimensions of 40 mm × 20 mm × 5 mm. The cladding powders, mix-up of Ni35A and TiC, were supplied with grain sizes of 48–106 μm. The microscopic profiles and chemical components are illustrated in Figure 1 and Table 1, respectively.

Table 1. Chemical components of Ni35A and TiC particle (wt.%).

Element	C	Si	O	Fe	Cr	B	T.C (Total Carbon)	F.C (Free Carbon)	N	Ni
Ni35A	0.32	3.35	<0.05	2.75	7.75	1.65	–	–	–	Rest
TiC	–	0.02	0.5	0.08	–	–	>18.8	<0.5	0.5	–

(a) (b)

Figure 1. SEM image of cladding powder. (**a**) Ni35A; (**b**) TiC.

The laser cladding system used in work, as depicted in Figure 2, was comprised of an IPG laser generator YLS-3000 (Burbach, Germany), the wavelength is 1064 nm, continuous mode, a Lasermesh FDH0273 laser head (Novi, MI, USA) with a focal length of 300 mm, a FANUCM-710iC/50 industrial robot (Yamanashi, Japan), a TongfeiTFLW-4000WDR-01-3385 laser chiller (Sanhe, China), a Songxing CR-PGF-D-2 coaxial powder feeding machine (Fuzhou, China), a MitsubishiPLC controller, and SX14-012PLUSE laser pulse control system (Burbach, Germany). The laser was focused on the substrate with a spot diameter of 4 mm, and the cladding powder was supplied by the shielding gas argon at a pressure of 0.5 MPa.

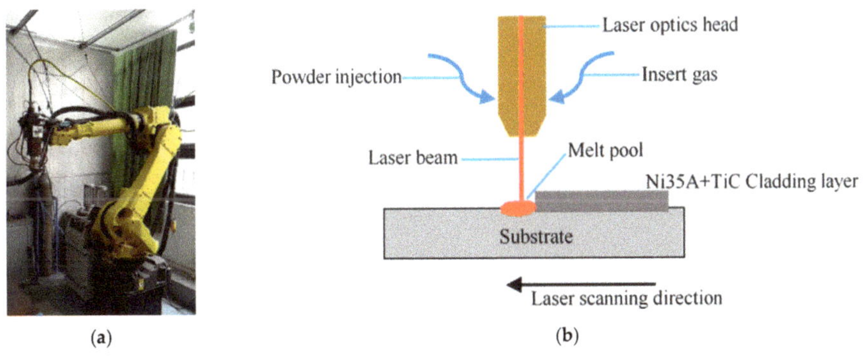

(a) (b)

Figure 2. Laser cladding system. (**a**) Experimental setup; (**b**) Schematic diagram.

Before cladding, the substrate was cleaned with acetone to remove surface impurities. The cladding powder was mixed up, according to the designated fraction, in a MITR-YXQM-2L planetary ball mill machine (MITR, Changsha, China) at a speed of 300 r/min for 30 min, followed by 30 min vacuum drying at a temperature of 120 °C. After laser cladding, the specimen was sectioned by wire-EDM (Electrical Discharge Machining), mounted, ground, polished, and etched in 4% nitric acid alcohol for 30 s. A series of observations was carried out, including microhardness test using MVA-402TS tester (HDNS, Shanghai, China) at a load of 500 gf and holding time of 30 s, macrographic measurement by a KH-1300 3D digital microscopy system (Hirox, Shanghai, China), metallurgical structure detection via a Hitachi TM3030 Plus scanning electron microscope (Hitachi, Tokyo, Japan), element quantification by A550I EDS (Austin, TX, USA), phase content analysis using XRD (Ultima IV, Rigaku Corporation, Tokyo, Japan) at a scanning speed of 4°/min and scanning angles range from 20° to 80°. The top surface of claddings was wear-tested using a Bruker UMT-2 universal tester (Bruker, Billerica, MA,

USA) at conditions illustrated in Table 2, and the wear volume was calculated after the observation of scratch profiles. The wear rate was computed using Equation (1) after obtaining the wear volume from scratch profiles:

$$r = \frac{\Delta V}{F \times d} \tag{1}$$

where ΔV is the wear volume of cladding, F is the applied load during the wear test, and d is the sliding distance.

Table 2. Friction and wear parameter table.

Wear Parameters	Unit	Specification
Friction pair	(mm)	Tungsten steel-ϕ 6 mm
Loading force	(N)	35
Speed	(mm/s)	10
Sliding distance	(mm)	4
Wear time	(min)	60
Mode	–	Linear reciprocating
Temperature	°C	Room temperature

Experimental runs were arranged according to the full factorial method using Design-expert 10.0. The experiment investigated the influences of four factors, including laser power, scanning speed, gas-powder flow rate, and particle ratio of TiC on the wear rate and microstructure evolution along the depth direction. The microstructure profile and element distribution in the cladding layer were also observed to reveal the correlation with macroscopic properties. The input variables and their levels are illustrated in Table 3. The experimental design matrix and observations are depicted in Table 4.

Table 3. Studied process variables and levels.

Variables	Notation	Unit	Levels of Input Variables		
			Code	**−1**	**1**
Laser power	*LP*	kW	Actual	1.2	1.4
Scanning speed	*SS*	mm/s		5	7
Gas flow	*GF*	L/h		1000	1400
TiC powder ratio	*PR*	wt.%		20	60

Table 4. Full factorial design and results of wear rate.

Run	*LP* (kW)	*SS* (mm/s)	*GF* (L/h)	*PR* (wt.%)	Wear Rate (μm³/N·mm) [1]	Micro-Hardness (HRC)
1	1.4	5	1000	20	41.981	63.4
2	1.4	7	1000	20	60.708	59.7
3	1.4	7	1000	60	18.997	72.6
4	1.2	7	1000	20	101.977	52.6
5	1.2	7	1400	20	96.205	53.3
6	1.4	7	1400	20	64.505	58.7
7	1.2	7	1400	60	21.157	72.0
8	1.4	5	1400	20	51.776	59.7
9	1.4	7	1400	60	17.202	73.4
10	1.2	5	1400	60	23.013	70.8
11	1.2	5	1000	60	25.586	70.0
12	1.2	5	1000	20	75.748	55.4
13	1.2	7	1000	60	41.821	69.9
14	1.2	5	1400	20	71.818	56.6
15	1.4	5	1000	60	11.922	75.2
16	1.4	5	1400	60	17.043	73.6

3. Results and Discussion

3.1. Subsection ANOVA of Wear Rate

Table 5 illustrates the result of ANOVA for wear rate. From the table, it can be seen that the linear correlation selected for wear rate had a p-value prob > F less than 0.0001, which denotes excellent goodness of fit. The influencing factors were selected to account for the response. The value of adequate precision was 22.514, much greater than 4, indicative of high resolution to input variations. The coefficient of determination R^2 was 0.9719, showing that the model fitted the experimental results well. The value of adjusted R^2 and predicted R^2 is 0.0297 [10,15], meaning that the established model had a good prediction capacity toward the input domain.

Table 5. Cladding efficiency variance analysis table.

Source	Sum of Squares	Degree of Freedom	Mean Square	F Value	p-Value Prob > F	
Model	12,586.60	5	2517.32	69.28	<0.0001	Significant
LP	1874.70	1	1874.70	51.59	<0.0001	
SS	671.91	1	671.91	18.49	0.0016	
PR	9407.88	1	9407.88	258.91	<0.0001	
$LP \times PR$	403.66	1	403.66	11.11	0.0076	
$SS \times PR$	228.46	1	228.46	6.29	0.0310	
Residual	363.36	10	36.34	–	–	
Cor total	12,949.97	15	–	–	–	
R-squared		0.9719	Adj R-squared		0.9579	
Pred R-squared		0.9282	Adeq precision		22.514	

From Table 5, it is apparent that the most significant factors influencing wear rate of the cladding layer were laser power and weight fraction of TiC (p-value < 0.001), followed by scanning speed. After ANOVA, the correlation between wear rate and process parameters was obtained, as depicted in Equation (2):

$$\begin{aligned}\text{Wear Rate} &= 289.923 - 208.70062 \times LP + 14.03769 \times SS - 3.34365 \times PR \\ &+2.251141 \times LP \times PR - 0.18893 \times SS \times PR\end{aligned} \quad (2)$$

3.2. Wear Rate Model

Figure 3 illustrates the distribution of residuals and the comparison of model-predicted values to experimental results. From the figure, it can be seen that the residuals of 16 experimental runs were along a straight line, which indicated normal distribution and thus random error for lack of fit. The comparison data were approximate to the line $y = x$, showing the little divergence of model predictions from experimental results. Therefore, the model for wear rate was established with good prediction accuracy.

Figure 4 plots the 3D response surface and corresponding contour of wear rate with respect to laser power and particle ratio of TiC. From the figure, it is clear that the wear rate of the cladding layer decreased with an increase in laser power and weight fraction of TiC. Further investigation of phase composition, microstructure, and element distribution is required to understand the mechanism of how process parameters affect the wear rate. Hardness and wear resistance were positively correlated under the same friction situation.

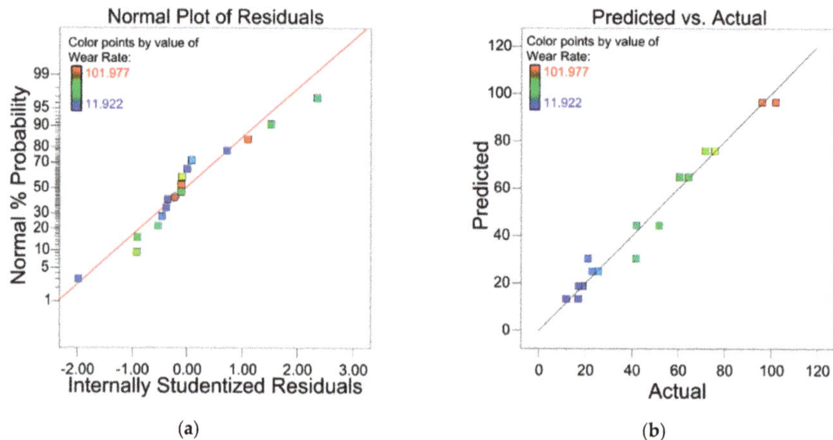

(**a**) (**b**)

Figure 3. (**a**) Distribution of residual wear rate, (**b**) Distribution of predicted and actual wear rate.

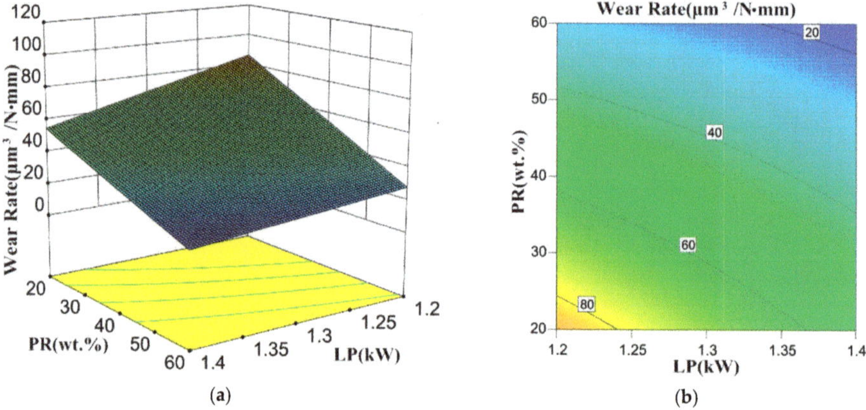

(**a**) (**b**)

Figure 4. The response of wear rate to significant factors *LP* and *PR*. (**a**) 3D response surface; (**b**) contour plot.

3.3. Phase Analysis

Figure 5 shows the phase analysis of the wear test on the cladding surface by the XRD test. It was found that the cladding layer was mainly composed of $TiC_{0.957}$(PDF#89-2716) and Ni_3Fe(PDF#88-1715). The diffraction spectrum for different process conditions varied from each other only in the relative intensity of peaks. This was mainly due to the weight fraction of TiC in cladding powder, and the grain growth status occurred by process parameters. Specifically, for the 11# specimen, where TiC particles amount to 60% weight of deposited powder, the peak intensity of $TiC_{0.957}$ was significantly strengthened as expected. It was also observed that the percentage of element C in TiC composition had decreased. This would be caused by the element loss during the heating process of laser-material interaction and replacement with Ni in the crystal structure of TiC at the molten pool [16].

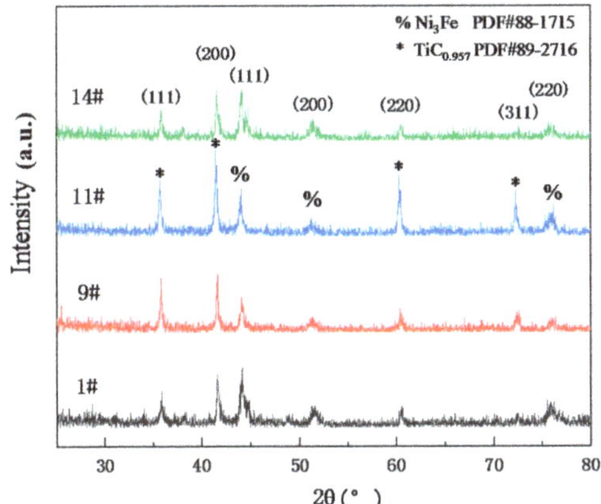

Figure 5. XRD spectrum.

3.4. Microstructure and Element Distribution

Figure 6 illustrates the element distribution of cladding layers at laser powers of 1.2 kW and 1.4 kW (Scanning speed of 7 mm/s, the gas-powder flow rate of 1000 L/h, and 60% of TiC). From the figure, it can be found the molten pool experienced strong convective flow during laser cladding. With higher laser input, the molten pool was prolonged, the element Fe was more diffused, by convection and diffusion, into the cladding layer from the substrate, especially to the boundary of three phases cladding layer-substrate-air and center of the molten pool. This, to a certain extent, diluted the cladding layer and decreased the fraction of the hard phase, thereby deteriorating the wear resistance.

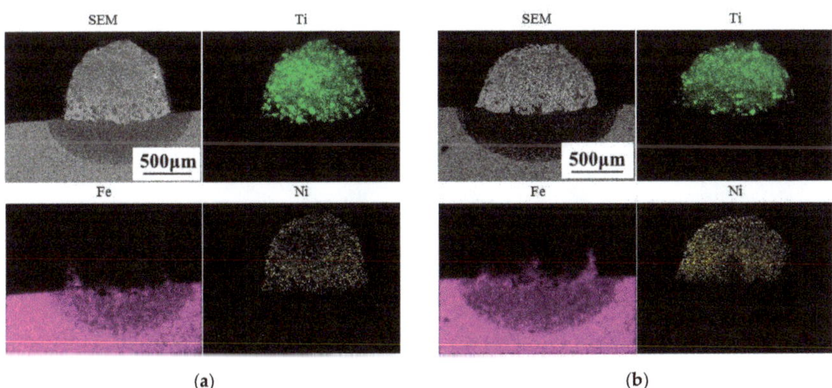

Figure 6. Element distribution of cladding layers at different laser powers. (**a**) 13# sample: 1.2 kW, 7 mm/s, 1000 L/h, 60% TiC; (**b**) 3# sample: 1.4 kW, 7 mm/s, 1000 L/h, 60% TiC.

Figure 7 compares the microstructure of cladding layers at different laser powers. It can be seen that a large amount of dendritic crystal existed in the top, center, and bottom of the cladding layer. Element analysis by EDS showed that the fraction of TiC at the dark area was obviously greater than that of light area, where the percentage of Ni overweighed. Considering the XRD result, the microstructures in the dark and light areas were probably TiC and Ni-based alloy, respectively. The dendritic microstructure is probably produced from the growth of renucleared TiC grain after decomposition due to heating at the laser cladding process [17].

As the laser power increased, more energy input was imposed onto the cladding powder. The more hard phase was thus decomposed, resulting in a decrease of hard phase retaining in the cladding layer. The microstructure of Ni-based alloy at different laser powers was a flatten crystal. At higher laser power (1.4 kW), the grain size was greater with more apparent crystal boundary. This was because increased laser power had prolonged the span of the molten pool, increased thermal gradient, and enhanced heat convection. Part of Fe elements and Ti elements left by the decomposition of TiC was concentrated in the grain boundaries, and this grain boundary structure with relatively weak mechanical properties will further reduce the inter-grain bonding strength and increase the wear rate [18].

According to the literature [5,19], the microstructure of the cladding layer is mainly determined by the solidification rate of the molten pool. At a certain scanning speed, the ratio of G to R (G is the temperature gradient, and R is the solidification rate) declined from the bottom of the molten pool to the top, while the growth rate of crystals at the same cross-section increased. As illustrated by the microstructure in the middle area in Figure 7a,b, due to the variation of G/R, the grain size of dendritic TiC was much greater than that of the bottom. The microstructure at the molten pool top manifested similar circumstances as the top and middle areas shown in Figure 7a,b. The dark grey hard phase at the top region of Figure 7a (2# sample) accounted for 16.84%, and the dark gray hard phase at the top of Figure 7b (4# sample) accounted for 15.29%. As shown in Table 4, the hardness of the 2# sample was 59.7 HRC, the hardness of the 4# sample was 52.6 HRC. The laser cladding hard particle-reinforced composite coating had high hardness, uniform structure, and excellent resistance to abrasive wear. When the laser power was 1.2 kW, it was found that some secondary dendrite axis was fully grown at the top area of the cladding layer, which facilitates anchoring in the Ni-based alloy and avoids hard phase being sloughed as abrasive grains [20]. Figure 8 shows the friction and wear diagram of the cladding layer with different laser powers. As shown in Table 6, the content of element Fe varied slightly through the top, middle, and bottom parts of the cladding layer.

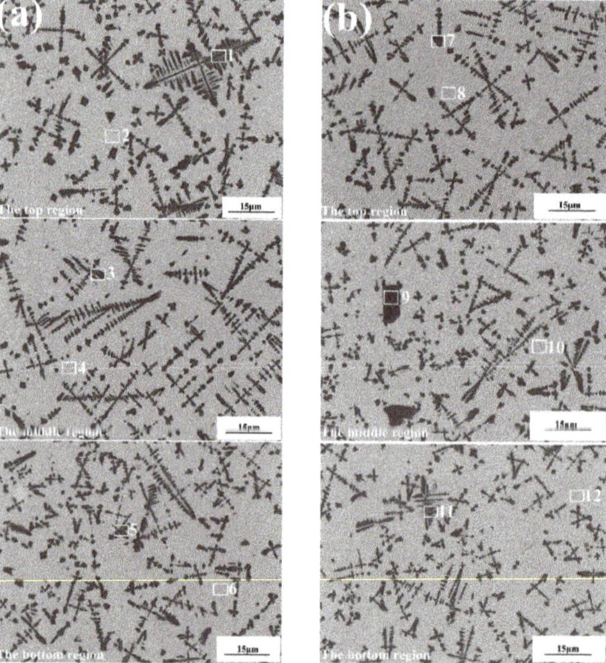

Figure 7. Microstructure of cladding layers at different laser powers: (**a**) 2# sample: 1.4 kW, 7 mm/s, 1000 L/h, 20% TiC; (**b**) 4# sample: 1.2 kW, 7 mm/s, 1000 L/h, 20% TiC.

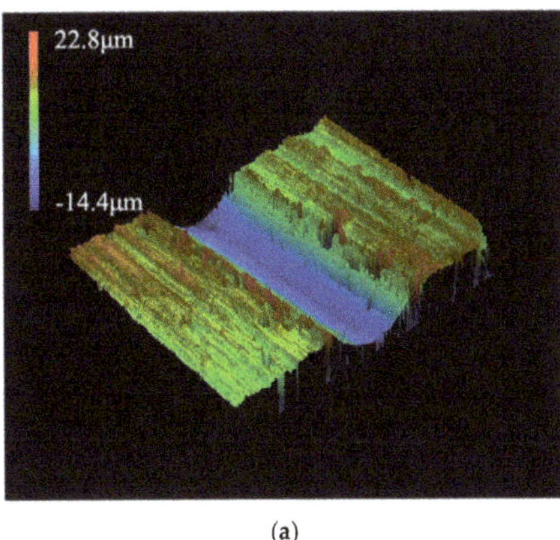

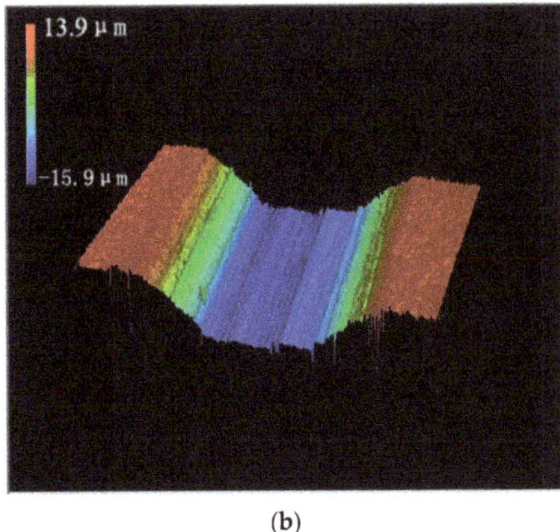

(a) (b)

Figure 8. Friction and wear diagram of cladding layer with different laser powers: (**a**) 2# sample: 1.4 kW, 7 mm/s, 1000 L/h, 20% TiC; (**b**) 4# sample: 1.2 kW, 7 mm/s, 1000 L/h, 20% TiC.

Table 6. The element content of cladding layers at different laser powers (Labels 1–6 belong to 2# sample, Labels 7–12 belong to 4# sample).

Label	C	Ti	Cr	Fe	Ni
1	13.265	53.329	5.071	3.709	24.626
2	8.389	4.742	5.947	6.616	74.296
3	14.109	52.713	5.959	4.021	23.198
4	9.604	4.763	4.583	5.388	75.662
5	12.301	54.728	5.175	8.842	19.224
6	6.906	3.084	5.864	15.929	68.216
7	12.057	48.635	5.753	6.740	26.816
8	8.074	4.114	5.487	12.220	70.106
9	11.725	57.388	6.568	5.705	18.614
10	7.524	3.993	5.022	12.844	70.617
11	10.507	58.422	7.186	8.441	15.444
12	5.303	3.268	6.430	15.957	69.042

The microstructure of the cladding layer obtained with 60% TiC indicated that the energy demand for decomposition and renuclearation of all the TiC particles grew increasingly with more TiC added in the cladding powder. The undecomposed TiC became nucleation sites and grew into the microstructure, as illustrated in Figure 9b. According to the distribution of thermal gradients along the cladding layer, TiC clustered more at the top and grew fully, while the TiC at the middle part grew with exceptional size, and at the bottom, more dendrite was obtained. The dark grey hard phase at the top region of Figure 9a (12# sample) accounted for 18.57%, and the dark gray hard phase at the top of Figure 9b (11# sample) accounted for 71.8%. As shown in Table 4, the hardness of the 12# sample was 55.4 HRC, and the hardness of the 11# sample was 70 HRC. From the energy spectrum by EDS, the percentage of TiC in the dark area increased obviously up to 70% as the weight fraction increased. Since TiC contained a more hard phase, the wear resistance was thus improved. With more TiC in the cladding powder, the effect of second phase enhancement became more apparent, which further strengthened the wear resistance. Figure 10 shows that the wear rate of the cladding layer decreased as the weight fraction of TiC increased in the cladding powder. This was due to the characteristic of high hardness and good wear resistance for the ceramic powder TiC. It was reported that the wear resistance of alloys was simply proportional to the area percentage of the hard phase. Increasing TiC percentage significantly improved the hardness and wear resistance of the cladding layer and declined the wear rate per unit time. As shown in Table 7, the content of element Fe varied slightly through the top, middle, and bottom parts of the cladding layer. This is owing to the fact that most of the imposed energy is absorbed by the TiC for decomposition and nucleation [21]. It also resulted from the decreased flowability of the melt pool as a more hard phase of the high melting point was mixed. Therefore, the diffusion of Fe was not widely observed, and the dilution effect was restrained.

Table 7. Element content table of micro TiC cladding layer structure determination. (Labels 1–6 belong to 12# sample, Labels 7–12 belong to 11# sample).

Label	C	Ti	Cr	Fe	Ni
1	14.889	44.969	7.795	2.346	30.001
2	9.437	4.940	5.295	3.143	77.185
3	10.135	54.201	5.813	2.501	27.350
4	6.102	5.536	5.931	3.222	79.209
5	10.017	58.422	7.372	6.254	17.935
6	6.163	3.833	5.462	6.435	78.107
7	13.114	75.288	1.410	1.402	8.787
8	8.546	13.885	4.114	2.605	70.849
9	11.899	72.929	2.185	2.194	10.794
10	6.867	13.367	5.460	3.961	70.344
11	11.245	71.978	1.311	5.997	9.740
12	6.218	9.564	5.820	9.964	68.433

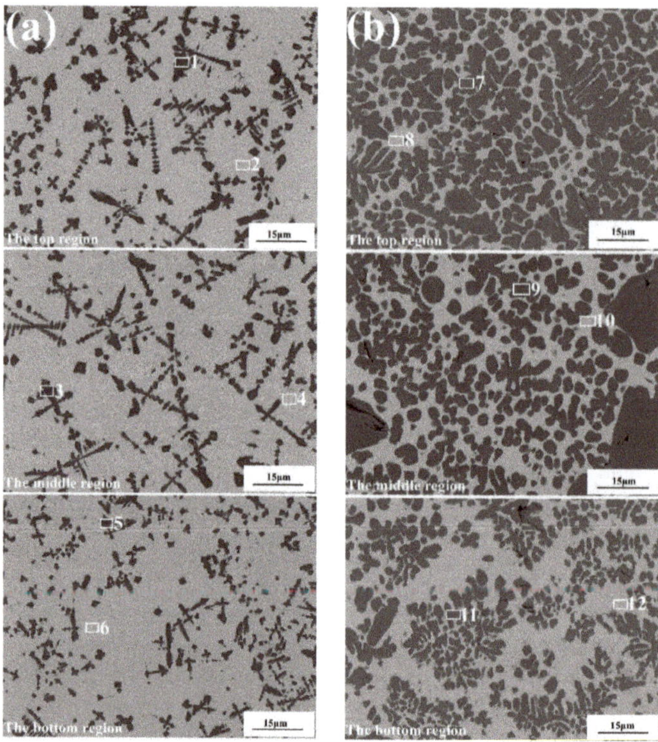

Figure 9. Microstructure of the top, bottom, and middle of the cladding layer with different TiC powder ratios: (**a**) 12# sample, 1.2 kW, 5 mm/s, 1000 L/h, 20% TiC; (**b**) 11# sample, 1.2 kW, 5 mm/s, 1000 L/h, 60% TiC.

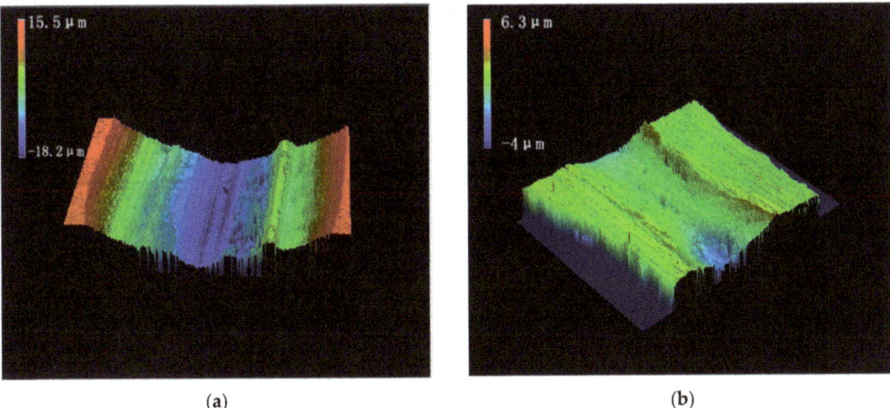

(a) (b)

Figure 10. Friction and wear diagram of cladding layer with different TiC powder ratios: (**a**) 12# sample: 1.2 kW, 5 mm/s, 1000 L/h, 20% TiC; (**b**) 11# sample: 1.2 kW, 5 mm/s, 1000 L/h, 60% TiC.

Figure 11 reflects the interaction effect of scanning speed and the weight ratio of TiC on the wear rate of the cladding layer. It can be seen that the wear rate increased as the scanning speed increased. This was because scanning speed determined the time span of the melt pool affected by the laser beam. The larger the speed, the shorter the energy effect time, which was adverse to thorough decomposition and renucleation of TiC. Specifically, at a higher scanning speed of 7 mm/s, as shown in Figure 11a, a large amount of fine TiC grain was observed at the top region of the cladding layer due to the more laser energy input. Compared with a needle or dendritic microstructure, near sphere structure was much preferable as it mitigated the possibility of stress concentration. The contact area between the spherical particles and the substrate was small, and the degree of fit between them was low, so the spherical particles were more likely to fall off during the friction process, resulting in abrasive wear, which reduced the wear resistance and increased the wear rate per unit time. As the scanning speed was decreased, the effect time of laser beam on the cladding powder and substrate was increased, and the undercooling was thus enlarged. The dendritic TiC crystal structure was obtained with finer sizes at the bottom. Furthermore, the smaller the speed, the larger amount the secondary phase at the crystal boundary. From the results of EDS, as listed in Table 8, the content of element Fe in Figure 12a was much greater than that in Figure 12b. The increase in Fe content, to some extent, diluted the fraction of the hard phase, thereby deteriorating the wear resistance. However, this was comparably less influencing the effect of the profile and distribution of the hard phase. It was presumed that the above-mentioned two effects counteract with each other so that the influence of scanning speed on the wear resistance is weakened even though the microstructure had been significantly changed [22]. The dark grey hard phase at the top region of Figure 12a (6# sample) accounted for 15.89%, and the dark gray hard phase at the top of Figure 12b (8# sample) accounted for 16.52%. As shown in Table 4, the hardness of the 6# sample was 58.7 HRC, and the hardness of the 8# sample was 59.7 HRC. Figure 13 shows the friction and wear diagram of the cladding layer at different scanning speeds.

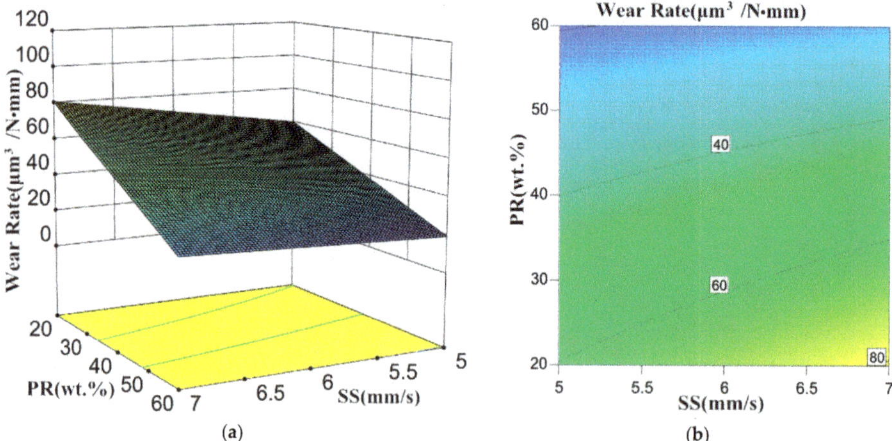

Figure 11. (**a**) 3D response curve of the interaction between *SS* and *PR* against wear rate; (**b**) Contour map of the interaction between SS and PR against wear rate.

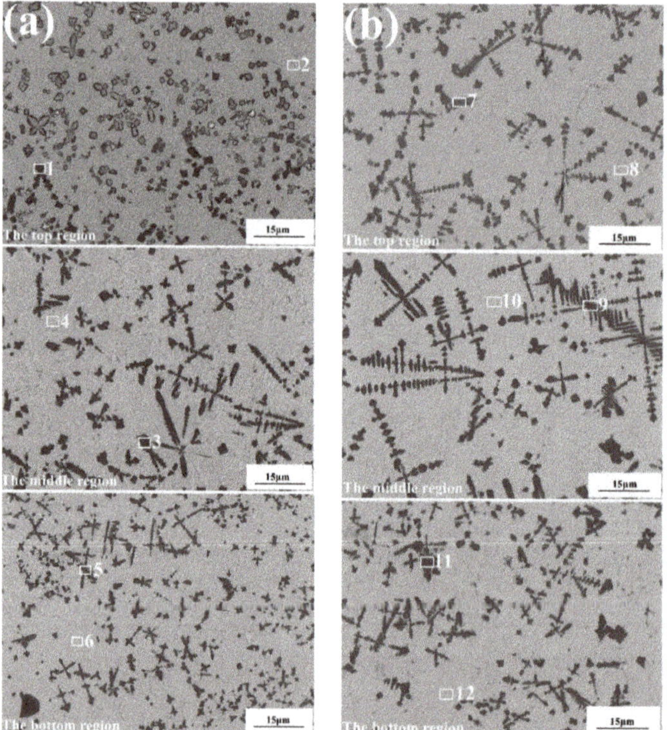

Figure 12. Microstructure of the top, middle, and bottom of the cladding layer at different scanning speeds: (**a**) 6# sample: 1.4 kW, 7 mm/s, 1400 L/h, 20% TiC; (**b**) 8# sample: 1.4 kW, 5 mm/s, 1400 L/h, 20% TiC.

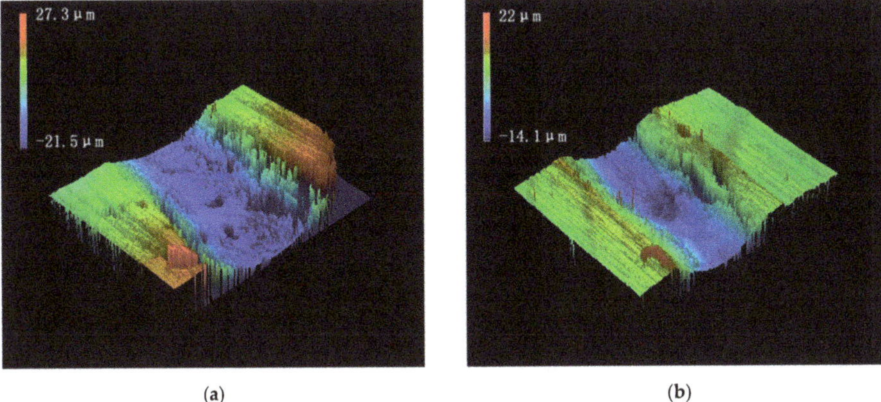

(a) (b)

Figure 13. Friction and wear diagram of cladding layer at different scanning speeds: (**a**) 6# sample: 1.4 kW, 7 mm/s, 1400 L/h, 20% TiC; (**b**) 8# sample: 1.4 kW, 5 mm/s, 1400 L/h, 20% TiC.

Table 8. Element content table of cladding structure determination at different scanning speeds. (Labels 1–6 belong to 6# sample, Labels 7–12 belong to 8# sample.)

Label	C	Ti	Cr	Fe	Ni
1	12.057	48.635	5.753	6.740	26.816
2	8.074	4.114	5.487	12.220	70.106
3	11.725	57.388	6.568	5.705	18.614
4	7.524	3.993	5.022	12.844	70.617
5	10.507	58.422	7.186	8.441	15.444
6	5.303	3.268	6.430	15.957	69.042
7	13.265	53.329	5.071	3.709	24.626
8	8.389	4.742	5.947	6.616	74.296
9	14.109	52.713	5.959	4.021	23.198
10	9.604	4.763	4.583	5.388	75.662
11	12.301	54.728	5.175	8.842	19.224
12	6.906	3.084	5.864	15.929	68.216

4. Conclusions

To investigate the influences of process parameters and powder formulation on microstructure, element distribution, and wear behavior, a series of laser cladding experimental runs was conducted on the AISI 1045 mild carbon steel with the cladding powder of Ni35A/TiC based on full factorial design. The correlation between process parameters and the wear resistance was then established after ANOVA of experimental results. Based on the results, the following conclusions can be obtained:

- Based on ANOVA, the established response model for the wear rate of the cladding layer has good prediction accuracy with R^2 of 0.9719. The most significant factors influencing the wear rate are laser power and powder formulation.
- The main phases of the cladding layer are Ni_3Fe and $TiC_{0.957}$. A small amount of C overflows the melt pool of TiC after solidus solution or decomposition. For 20% TiC, the cladding layer is mainly composed of one axis dendrite, while for 60% TiC, the grain of hard phase grows up completely without preferable directions, avoiding stress concentration at sharp tips and improving the mechanical properties.
- Higher laser power would augment the convection-diffusion of the melt pool. From the element distribution by EDS, the element Fe diffuses from the substrate to the cladding layer along the direction of convection, thus increasing the dilution of Fe to the cladding layer and wear volume.

- The microstructure at the top region of the cladding layer can be controlled by scanning speed. At lower speed (5 mm/s), the dendritic crystal of TiC transforms into near sphere fine grains for the top area. This refines the microstructure to reduce stress concentration at grain tips and improves the wear properties.

Author Contributions: Methodology, X.H. and C.L.; experiment, C.L.; analysis, X.H., C.L., H.Z., C.C., G.L., J.J., M.F., and M.Z.; writing—original draft preparation, X.H. and C.L.; writing—review and editing, C.C. and H.Z.; supervision, X.H. and C.C. All authors have read and agreed to the published version of the manuscript.

Funding: This research was funded by Fujian University of Technology, Grant No. GY-Z18163; Fujian Innovation Center of Additive Manufacturing, Grant No. ZCZZ20-04; and Fujian Provincial Department of Human Resources and Social Affairs, Grant No. GY-Z19045.

Acknowledgments: The authors gratefully acknowledge the support from the Public Service Platform for Technical Innovation of Machine Tool Industry in Fujian Province at the Fujian University of Technology.

Conflicts of Interest: The authors declare no conflict of interest.

References

1. Wang, K.; Chang, B.; Chen, J.; Fu, H.; Lin, Y.; Lei, Y. Effect of molybdenum on the microstructures and properties of stainless steel coatings by laser cladding. *Appl. Sci.* **2017**, *7*, 1065. [CrossRef]
2. Zhang, M.; Zhou, X.; Yu, X.; Li, J. Synthesis and characterization of refractory TiZrNbWMo high-entropy alloy coating by laser cladding. *Surf. Coat. Technol.* **2017**, *311*, 321–329. [CrossRef]
3. Liu, J.; Yu, H.; Chen, C.; Weng, F.; Dai, J. Research and development status of laser cladding on magnesium alloys: A review. *Opt. Laser Eng.* **2017**, *93*, 195–210. [CrossRef]
4. Lv, Y.H.; Li, J.; Tao, Y.F.; Hu, L.F. High-temperature wear and oxidation behaviors of TiNi/Ti_2Ni matrix composite coatings with TaC addition prepared on Ti6Al4V by laser cladding. *Appl. Surf. Sci.* **2017**, *402*, 478–494. [CrossRef]
5. Lian, G.; Zhang, H.; Zhang, Y.; Tanaka, M.L.; Chen, C.; Jiang, J. Optimizing processing parameters for multi-track laser cladding utilizing multi-response grey relational analysis. *Coatings* **2019**, *9*, 356. [CrossRef]
6. Sabahi, N.A.; Ahmadi, Z.; Babapoor, A.; Shokouhimehr, M.; Shahedi Asl, M. Microstructure and thermomechanical characteristics of spark plasma sintered TiC ceramics doped with nano-sized WC. *Ceram. Int.* **2019**, *45*, 2153–2160. [CrossRef]
7. AlMangour, B.; Grzesiak, D.; Yang, J. In situ formation of TiC-particle-reinforced stainless steel matrix nanocomposites during ball milling: Feedstock powder preparation for selective laser melting at various energy densities. *Powder Technol.* **2018**, *326*, 467–478. [CrossRef]
8. Wang, C.; Gao, Y.; Zeng, Z.; Fu, Y. Effect of rare-earth on friction and wear properties of laser cladding Ni-based coatings on 6063Al. *J. Alloys Compd.* **2017**, *727*, 278–285. [CrossRef]
9. Dudziak, T.; Boron, L.; Gupta, A.; Saraf, S.; Skierski, P.; Seal, S.; Sobczak, N.; Purgert, R. Steam oxidation resistance and performance of newly developed coatings for Haynes® 282® Ni-based alloy. *Corros. Sci.* **2018**, *138*, 326–339. [CrossRef]
10. Lian, G.; Zhang, H.; Zhang, Y.; Yao, M.; Huang, X.; Chen, C. Computational and experimental investigation of micro-hardness and wear resistance of Ni-based alloy and TiC composite coating obtained by laser cladding. *Materials* **2019**, *12*, 793. [CrossRef]
11. Hong, C.; Gu, D.; Dai, D.; Alkhayat, M.; Urban, W.; Yuan, P.; Cao, S.; Gasser, A.; Weisheit, A.; Kelbassa, I.; et al. Laser additive manufacturing of ultrafine TiC particle reinforced Inconel 625 based composite parts: Tailored microstructures and enhanced performance. *Mater. Sci. Eng. A* **2015**, *635*, 118–128. [CrossRef]
12. Xu, X.; Mi, G.; Xiong, L.; Jiang, P.; Shao, X.; Wang, C. Morphologies, microstructures and properties of TiC particle reinforced Inconel 625 coatings obtained by laser cladding with wire. *J. Alloy. Compd.* **2018**, *740*, 16–27. [CrossRef]
13. Saroj, S.; Sahoo, C.K.; Tijo, D.; Kumar, K.; Masanta, M. Sliding abrasive wear characteristic of TIG cladded TiC reinforced Inconel825 composite coating. *Int. J. Refract. Met. Hard Mater.* **2017**, *69*, 119–130. [CrossRef]
14. Sahoo, C.K.; Masanta, M. Microstructure and mechanical properties of TiC-Ni coating on AISI304 steel produced by TIG cladding process. *J. Mater. Process. Technol.* **2017**, *240*, 126–137. [CrossRef]

15. Olakanmi, E.O.; Nyadongo, S.T.; Malikongwa, K.; Lawal, S.A.; Botes, A.; Pityana, S.L. Multi-variable optimisation of the quality characteristics of fiber-laser cladded Inconel-625 composite coatings. *Surf. Coatings Technol.* **2019**, *357*, 289–303. [CrossRef]
16. Zhao, Y.; Yu, T.; Sun, J.; Jiang, S. Microstructure and properties of laser cladded B4C/TiC/Ni-based composite coating. *Int. J. Refract. Met. Hard Mater.* **2020**, *86*, 105112. [CrossRef]
17. He, X.; Song, R.G.; Kong, D.J. Effects of TiC on the microstructure and properties of TiC/TiAl composite coating prepared by laser cladding. *Opt. Laser Technol.* **2019**, *112*, 339–348. [CrossRef]
18. Cao, Y.; Zhi, S.; Qi, H.; Zhang, Y.; Qin, C.; Yang, S. Evolution behavior of ex-situ NbC and properties of Fe-based laser clad coating. *Opt. Laser Technol.* **2020**, *124*, 105999. [CrossRef]
19. Liu, C.; Li, C.; Zhang, Z.; Sun, S.; Zeng, M.; Wang, F.; Guo, Y.; Wang, J. Modeling of thermal behavior and microstructure evolution during laser cladding of AlSi10Mg alloys. *Opt. Laser Technol.* **2020**, *123*, 105926. [CrossRef]
20. Liang, J.; Liu, Y.; Li, J.; Zhou, Y.; Sun, X. Epitaxial growth and oxidation behavior of an overlay coating on a Ni-base single-crystal superalloy by laser cladding. *J. Mater. Sci. Technol.* **2019**, *35*, 344–350. [CrossRef]
21. Li, Y.; Dong, S.; Yan, S.; Li, E.; Liu, X.; He, P.; Xu, B. Deep pit repairing of nodular cast iron by laser cladding NiCu/Fe-36Ni low-expansion composite alloy. *Mater. Charact.* **2019**, *151*, 273–279. [CrossRef]
22. Zhan, X.; Qi, C.; Gao, Z.; Tian, D.; Wang, Z. The influence of heat input on microstructure and porosity during laser cladding of Invar alloy. *Opt. Laser Technol.* **2019**, *113*, 453–461. [CrossRef]

Article

Effect of Interlayer Cooling on the Preparation of Ni-Based Coatings on Ductile Iron

Yuyu He [1,2,3], Yijian Liu [1,2,3,*], Jiquan Yang [1,2,3], Fei Xie [1,2,3], Wuyun Huang [1,2,3], Zhaowei Zhu [1,2,3], Jihong Cheng [1,2,3] and Jianping Shi [1,2,3,*]

1 School of Electrical and Automation Engineering, Nanjing Normal University, Nanjing 210023, China; 171802031@stu.njnu.edu.cn (Y.H.); 63047@njnu.edu.cn (J.Y.); xiefei@njnu.edu.cn (F.X.); 171835010@stu.njnu.edu.cn (W.H.); 181802030@stu.njnu.edu.cn (Z.Z.); 63029@njnu.edu.cn (J.C.)

2 Jiangsu Key Laboratory of 3D Printing Equipment and Manufacturing, Nanjing 210023, China

3 Nanjing Institute of Intelligent High-End Equipment Industry Co. Ltd., Nanjing 210023, China

* Correspondence: 61213@njnu.edu.cn (Y.L.); jpshi@njnu.edu.cn (J.S.); Tel.: +86-02585481135 (Y.L.); +86-02585481135 (J.S.)

Received: 25 March 2020; Accepted: 27 May 2020; Published: 4 June 2020

Abstract: In metal additive manufacturing without interlayer cooling, the macro-size of the layer itself is difficult to control due to the thermal storage effect. The effect of interlayer cooling was studied by cladding Ni-based coatings on the substrate of ductile iron. The results show that under the same process parameters, compared with non-interlayer cooling deposition, the dilution rate is better, and the thickness increase of interlayer cooling deposition is more uniform, which is conducive to controlling the macro-size of the interlayer cooling deposition. Furthermore, interlayer cooling deposition has fewer impurities and more uniform microstructures. Moreover, the average grain size is refined and the dendrite growth is inhibited, which improves the mechanical properties of the coating. Therefore, the hardness of the interlayer cooling specimens is greater than that of the non-interlayer-cooled specimens.

Keywords: metal additive manufacturing; nickel-based alloy; microstructure; cooling effect

1. Introduction

Ductile iron possesses the properties of high strength, toughness, wear resistance, shock absorption, easy cutting, notch insensitive, etc. However, due to the harsh working conditions, the surface wear of ductile iron can cause mechanical failure [1,2]. Surface modification technology is an effective way to enhance surface strengthen, overcome the limitations of shape and size, and reduce costs [3]. Due to its good properties and low cost, Ni-based powders are often used to improve the surface properties of substrates, such as strength, wear resistance, and corrosion resistance [4–6].

Mughal et al. [7,8] used the finite element method to study the process of metal additive manufacturing process. It has been proven that continuous deposition without interlayer cooling leads to high and uniform preheating of the substrate, which reduces the deformation of the cladding layer. However, at the same time, it was found that the continuous multilayer deposition causes a thermal storage effect, which results in the loss of control over dimensional tolerances.

Delinger et al. [9] studied the effect of interlayer cooling time on the distortion and residual stress of Ti–6Al–4V powder. The results show that a shorter interlayer cooling time produces lower distortion and residual stresses. Nevertheless, shortening the cooling time can excessively increase the energy input into the system. Overheating can consequently lead to undesirable remelting, poor surface finish, and poor dimensional tolerances in the final part.

Chen et al. [10] studied the effect of improving the base cooling effect on the metal additive manufacturing process. It is found that the crystal orientation of the specimens can be improved by imposing a continuous water flow on the back of the substrate. As described in the paper, due to the thermal storage during the deposition process, the microstructure of the deposition is uneven. This means that as the number of layers increases, the cooling effect becomes worse. Furthermore, it is only suitable for specimens with simple shapes, not for specimens with complex shapes.

At present, there are many studies on laser process parameters. Weng et al. [11] studied the effects of laser-specific energy on the microstructures and properties of the cladding layers. Liu et al. [12] compared the structural and mechanical properties of Inconel 718 prepared with or without argon protection during laser cladding. Cheng et al. [13] studied the effects of laser energy density and scanning speed on the properties of the cladding layer. However, few studies have been done on the effect of interlayer cooling. In this paper, the differences in the microstructure and properties of two Ni-based alloys deposited on ductile iron under interlayer cooling or non-interlayer cooling conditions were investigated. This work can improve the quality of the cladding layer, especially strength and hardness, and help control dimensional tolerances.

2. Experimental Details

The experimental system consists of a 4 kW semiconductor laser (LDM 4000-100, Laserline, Mülheim-Kärlich, Germany), a 6-axis KUKA robot (KR 16-2), a high precision powder feeder, a side powder feeding nozzle, and an inert gas (Ar) protection box.

The substrate was a piece of ductile iron (ISO 1038/JS/500-7) with a size of 100 mm × 50 mm × 15 mm. The powder used in the experiment was Ni-based self-fluxing alloy powder (Wall Colmonoy Corporation, Madison Heights, MI, USA). The chemical compositions of the ductile iron and the powder are listed in Tables 1 and 2, respectively.

Table 1. Chemical composition of ductile iron (mass fraction (%)).

C	Si	Mn	S	P	Mg	Re
3.55–3.85	2.34–2.86	<0.6	<0.025	<0.08	0.02–0.04	0.03–0.05

Table 2. Chemical composition of cladding materials (mass fraction (%)).

Al	B	C	Cr	Fe	Si	Ni
1.1	1.0	0.14	3.5	2.0	2.7	Rem

Before the experiment, the powder was dried in a vacuum at 100 °C for 2 h to remove water vapor adsorbed on the surface. The substrate's surface was polished with sandpaper (180–400 grit sandpaper) and cleaned with acetone. The main process parameters used in the experiments are as follows: laser power of 1.6–2.5 kW, scanning speed of 8 mm/s, powder feeding speed of 20 g/min, gas flow rate of 12 L/min, laser spot diameter of 5 mm, and deposition thickness of 0.2 mm. Table 3 shows the deposition parameters in the experiment. After each cladding layer was completed, each interlayer cooling specimen was cooled to room temperature (about 5 min), and each non-interlayer cooling specimen was cladded without waiting time. The cladding equipment and one of the samples are shown in Figure 1.

Table 3. Experimental parameters, thickness of layers, and depth of penetration.

No.	Laser Power (W)	Number of Cladding Layers	Does it Include Cooling Time	Thickness of Layers (mm)	Penetration (μm)/ Dilution Rate (%)
1	1600	10	No	9.52	156/1.55
2	1600	10	Yes	10.01	149/1.67
3	2000	8	No	6.25	630/9.16
4	2000	8	Yes	8.01	714/8.18
5	2500	8	No	5.88	564/8.62
6	2500	8	Yes	6.92	625/8.28

The height of the cladding layers was measured by a micrometer. The cross-section of the specimen was cut by a wire-cut electrical discharge machine (EDM, DK7763, CHENGHONG, Jiangsu, China) along a vertical laser path. Metallographic specimens were prepared by mechanical polishing. The polished specimens were etched by a solution (200 mL H_2O + 200 mL HCl + 40 g $CuSO_4$). The heat-affected zone (HAZ) was etched by a solution (HF and HNO_3 1:1 mixing). The microstructures of the powder and the layers were analyzed by optical microscopy (OM, Axio Vert.A1, ZEISS, Oberkochen, Germany), and a scanning electron microscopy (SEM, SIGMA 04-03, ZEISS, Oberkochen, Germany) equipped with an energy-dispersive spectroscopy (EDS). The penetration data in Table 3, which indicate the distance from the deepest position of the substrate's melting part to the surface of the substrate, as schematically illustrated in Figure 2, were also measured by an optical microscopy. Meanwhile, the dilution ratio in Figure 2 is the ratio of penetration to thickness of layers. Phase constitution was identified by X-ray diffraction (XRD, D/MAX 2500/PC, Neo-confucianism, Tokyo, Japan). The microhardness of the single tracks was measured with a HV-5ACL microhardness tester (HENGYI, Shanghai, China) and a Vickers pyramidal-shaped diamond indenter under the load of 0.5 kg for 15 s of loading.

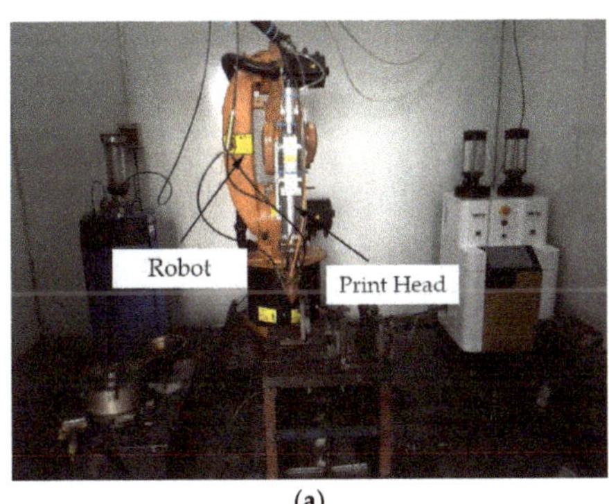

(a)

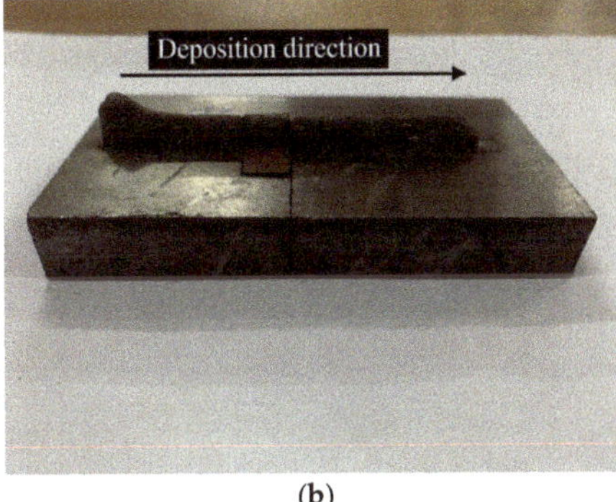

(b)

Figure 1. Pictures of the cladding equipment and specimens: (**a**) cladding equipment; (**b**) picture of one specimen.

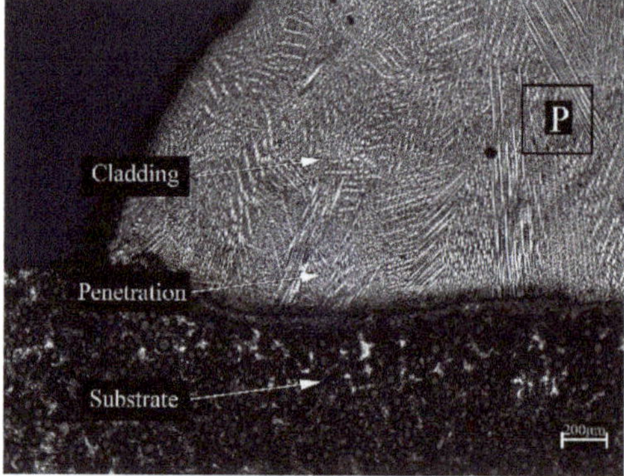

Figure 2. Overview of the cladding.

3. Results and Discussion

Figure 3a shows that the powder particles are regular spheres of 50–100 μm, which helps the smooth powder transfer during the deposition process. Figure 3b (the XRD pattern of the powder) shows that the main phase of the powder is γ-Ni.

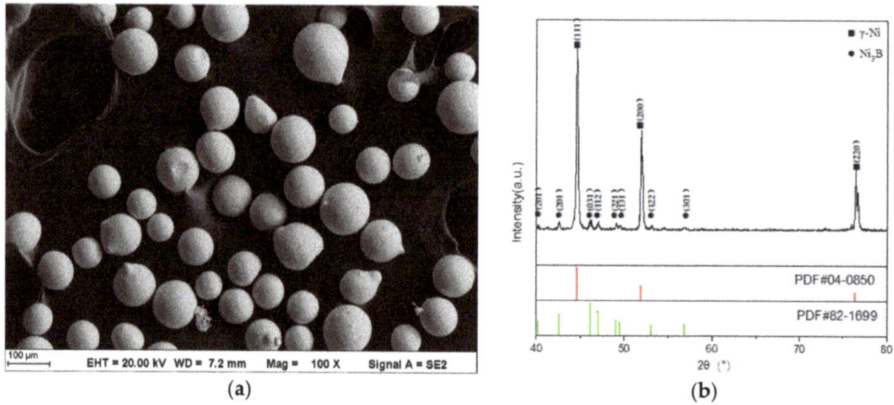

Figure 3. SEM image and XRD patterns of the powder: (**a**) SEM image of powder shape and size and (**b**) XRD pattern of the powder.

Figure 2 shows a schematic overview image of the cladding. There are no pores or cracks in the coating. Planar growth at the interface between the coating/substrate indicates good metallurgical bonding. The micromorphologies of the six specimens in Table 3 are shown in Figure 4. Their corresponding areas are in the middle of the samples (Area P in Figure 2). There are some bright and black impurities in coatings, as shown in Figures 4 and 5. Compared to the non-interlayer cooling specimens, the interlayer cooling specimens have less impurities. Through SEM and EDS results (Figure 6), the contents of Fe and Al in the bright impurities are 6.8% and 9.4%, respectively, which are higher than the normal values. Meanwhile, the contents of O and Cl are more than 20% in the black impurities. This may be due to the reaction between the black impurities and HCl during sample preparation. Hence, the bright impurities in Figure 4c may be the intermetallic compounds of Al, and the black impurities in Figure 4a may be oxides.

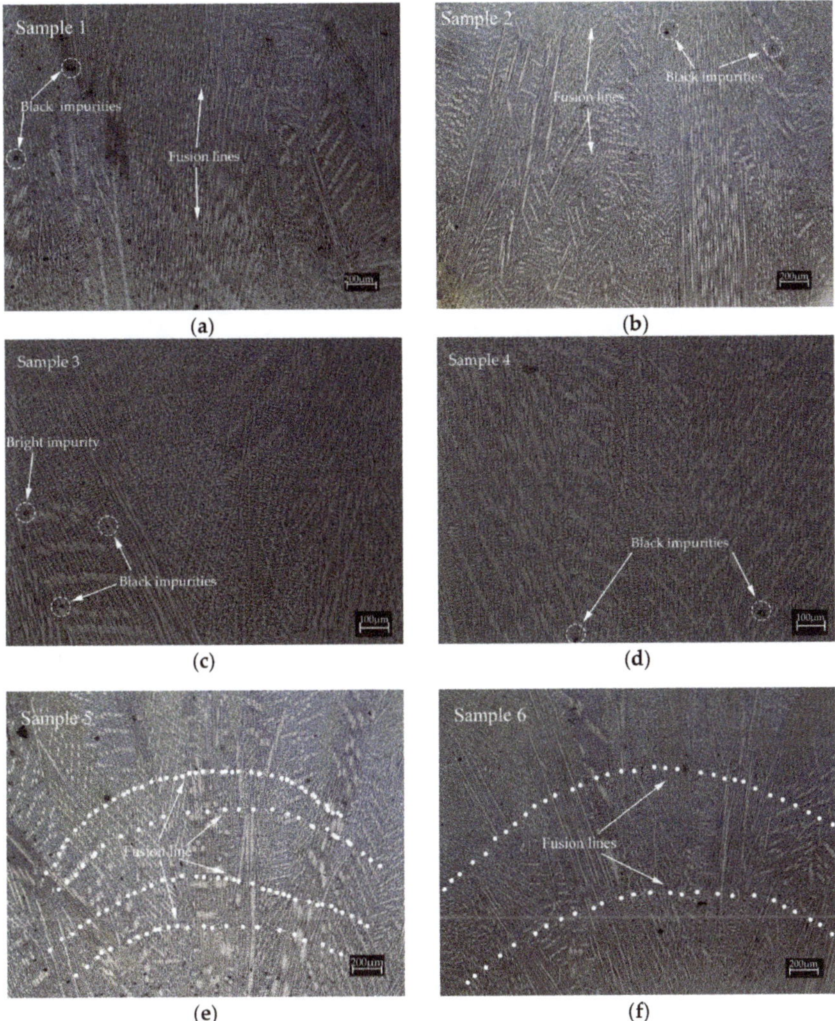

Figure 4. Optical microscopy (OM) images of 6 groups about fusion lines and impurities: (**a**) fusion lines and black impurities in Sample 1; (**b**) fusion lines and black impurities in Sample 2; (**c**) bright and black impurities in Sample 3; (**d**) black impurities in Sample 4; (**e**) fusion lines in Sample 5; and (**f**) fusion lines in Sample 6.

During the deposition process, the pre-deposited layer acts as a substrate. At the bottom of the molten pool, the low-melting eutectic compounds in the interdendritic region were remelted; then, they diffused into the molten pool and diluted the composition of the interdendritic liquid [14]. After that, the interdendritic liquid was restored to a composition suitable for dendritic growth. As a result, the un-remelted dendrite arms grew slightly larger than the dendrites in the surrounding area, which produced metallographical details of the fusion lines in the optical micrograph, as shown in Figure 4a,b. The fusion of each layer in Sample 5 and Sample 6 is highlighted by the white dashed lines in Figure 4e,f, respectively. Figure 7 shows a further enlarged detail of the fusion lines, which clearly shows the fusion lines produced by remelting. Due to the high energy input of laser, according to the metallographic diagram, part of the pre-deposited layer was remelted into the lower layer during the deposition of each layer. Therefore, the size of the molten pool is increased.

Figure 5. SEM images of the impurities: (**a**) bright impurity; (**b**) black impurity.

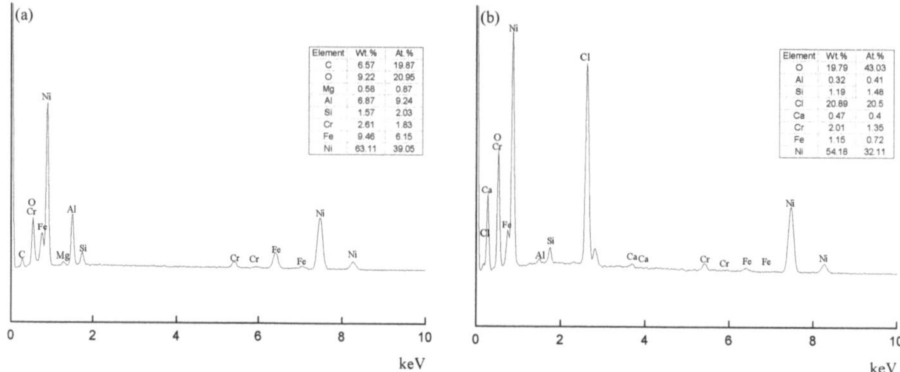

Figure 6. Energy-dispersive spectroscopy (EDS) results for the bright impurities (**a**) and black XRD impurities (**b**).

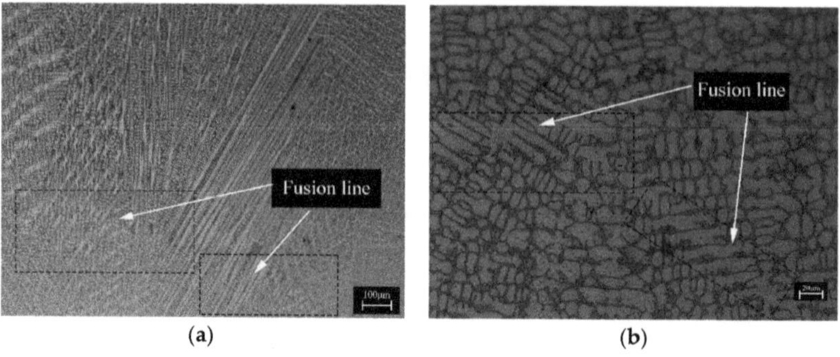

Figure 7. The detail of fusion line at 100× magnification (**a**) and at 400× magnification (**b**).

Compared with the interlayer cooling specimens, non-interlayer cooling deposition was preheated excessively due to heat storage effect, which caused the temperature of the molten pool to rise. Thus, the control of the dilution rate and the dimensional tolerance of the non-interlayer cooling deposition become harder [6,15,16]. Moreover, excessive preheating makes each layer of the non-interlayer cooling specimen thinner and the final layer thicker. As shown in Table 3, with the same process parameters, the thickness of the layers without cooling is slightly smaller than that with cooling. This is because the poor thermal conductivity of nickel causes heat buildup in the layer, which increases the diameter of the molten pool. The higher the temperature of the pre-deposited layer, the larger the molten pool will be. The larger size of the molten pool will form a cladding layer with a smaller thickness and a larger width. By comparing the specimens with and without interlayer cooling, the interlayer cooling specimens have a greater penetration depth but a smaller dilution rate. Sample 1 and Sample 2 were not considered because their penetration depths were too small to be representative. In order to ensure the high performance of the cladding layer, it is generally considered that the dilution rate should be less than 10%, preferably about 5% [17]. Furthermore, the direction of dendrite growth in the non-interlayer cooling specimens is more chaotic. This is due to the preheating of the substrate, which makes it difficult to dissipate heat, thereby making the dendrites grow disorderly.

Figure 8 shows high magnification micrographs of Sample 5 and Sample 6. Due to the high cooling rate, the typical rapid directional solidified structures composed of dendritic and interdendritic regions are shown. In contrast, the number of primary dendrites and secondary dendrites in the cladding specimens with interlayer cooling was significantly higher than that in the specimen without cooling. The smaller the number of dendrites per unit area, the greater the distance between the dendrites. According to the solidification theory, the dendrite spacing depends on the heat dissipation condition at the solidification interface. The stronger the heat dissipation ability, the smaller the influence range of the latent heat of the crystals precipitated from each branch, and the smaller the dendrite spacing. When directional solidification occurs, the cooling rate is determined by the heat dissipation capacity at the solid–liquid interface. The greater the cooling rate, the stronger the heat dissipation capability at the solid–liquid interface. Therefore, a high solidification rate often leads to fine dendrite arrangement. Interlayer cooling facilitates heat dissipation during deposition and provides faster cooling rates. Some studies have shown that at higher cooling rates, dendrite coarsening is inhibited, dendrite spacing decreases, nucleation rate increases, and the grain refinement and interfacial base increase [18,19], which prevent dislocation movement, balance plastic deformation, and increase the strength of the coating [20–22].

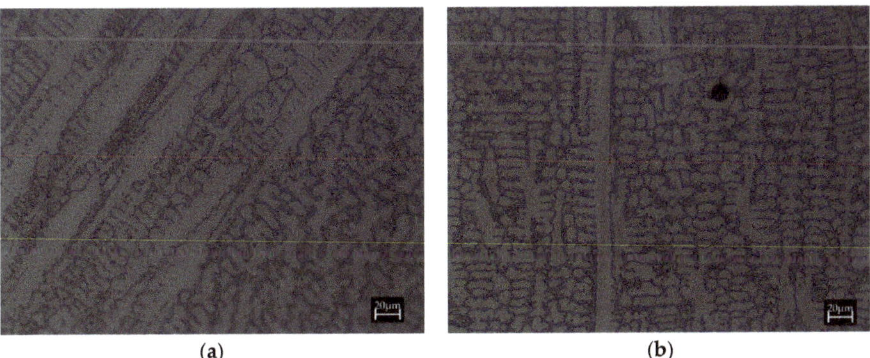

(a) (b)

Figure 8. Dendritic images of Sample 5 (**a**) and Sample 6 (**b**).

Since the constituent phases cannot be identified by optical microscopy, further SEM and EDS analyses were performed on the cladding layers. Figure 9 is a representative SEM image of Sample 5. Typical dendritic structures, including dendritic and interdendritic regions, can be clearly seen in the figure. Combine with the EDS results in Figure 10, it is judged that area A is the γ phase and area B is the γ' phase.

Figure 9. SEM image of cross-section morphology of Sample 5 coating.

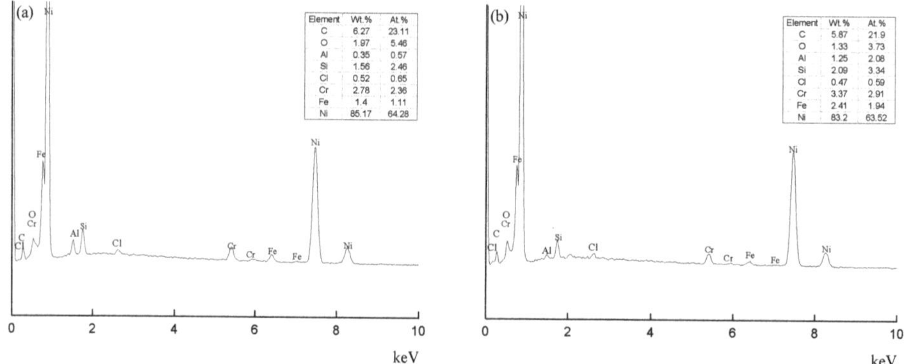

Figure 10. EDS results for area A (**a**) and area B (**b**).

To further determine the composition, six sets of specimens were analyzed by XRD, as shown in Figure 11. Sharp and broad peaks were observed, which corresponded to the γ and γ' phase of the Ni-base solid solution. Ni_3B phase and some other compounds of Ni were observed in Samples 1 and 2. Combined with the XRD pattern of the powder (Figure 3b), it may be due to insufficient laser power, resulting in more Ni3B residue powder in the coatings. At the same time, NiC_x was created. From the XRD results, no significant intermetallic oxides were observed due to the low volume fraction.

The grain sizes of in Figure 11 were calculated by Scheler's formula. The grain sizes of samples 1–6 are 33.2, 28.7, 56.7, 36.9, more than 100 and 39.5 nm, respectively. By comparing Samples 1, 3, and 5 or Samples 2, 4, and 6, we can find that the grain sizes of the sample increase with the increase of laser power. This is due to the temperature of molten pool increasing with the increase of laser power. However, there is no obvious difference in the increase of grain sizes of the samples with interlayer cooling, only from 28.7 increase to 39.5 nm, while the grain sizes of non-interlayer cooling change greatly, from 33.2 to more than 100 nm.

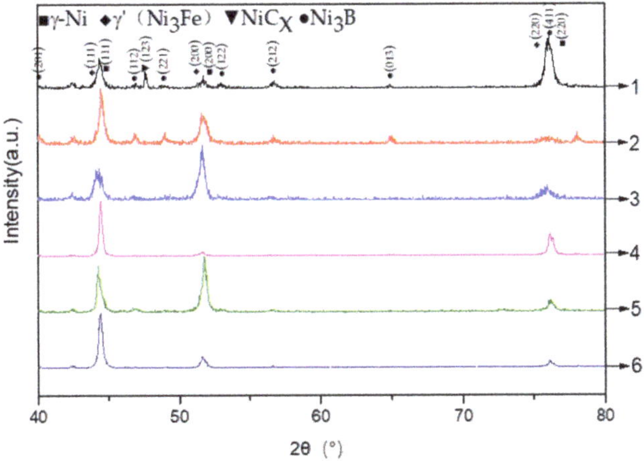

Figure 11. Diffraction patterns of 6 groups of single tracks.

Furthermore, by comparing Sample 1 and 2, there is no significant difference between the grain sizes of them, which shows that the effect of interlayer cooling on the grain size is small at the laser power of 1.6 kW. Nevertheless, with the increase of laser power, the difference of grain size becomes more and more significant. Especially when the laser power is 2.5 kW, the grain size of the samples without interlayer cooling is more than twice that of the samples with interlayer cooling. This shows that the effect of interlayer cooling will be improved with the increase of laser power.

The microhardness distribution from the top of the coatings to the substrate are shown in Figure 11. It indicates that the average microhardness of the cladding layer increases from 263 ± 8.5 $HV_{0.5}$ to 335 ± 6.9 $HV_{0.5}$. Under the same process parameters, the average microhardness of the specimens with and without interlayer cooling does not change significantly near the surface of the substrate due to the interaction between the cladding layer and the substrate. In Figure 12a, there is no significant difference in microhardness between Samples 1 and 2, especially near the surface of the substrate. This is due to the lower laser power and insufficient thermal input. As a result, the remelting area of the pre-deposited layer is less. In this area, the interlayer cooling effect is not significant, and the lower the power, the larger the area. As the distance increases, the average microhardness of the specimens with interlayer cooling begins to be higher than that of the specimen without interlayer cooling. By comparing the microhardness of the two samples in Figure 12a–c, we can find that the microhardness difference is more and more significant with the increase of laser power. The microhardness of the same thickness exceeds 3–4% in Figure 12b, while that exceeds 6–8% in Figure 12c. This is due to the poor thermal conductivity of Ni-based metals. As deposits accumulate, the heat dissipation effect decreases gradually, resulting in heat accumulation and relative preheating of the substrate. The better heat dissipation makes the grain size of the interlayer cooling specimens relatively fine and uniform and increases the total area of the grain boundaries, thereby preventing dislocation movement, balancing plastic deformation, and improving the strength of the coating. When the laser power is low (Figure 12a), the influence is insignificant. With the increase of laser power, there is a significant difference between the two grain sizes, leading to a significant difference between the two conditions. This shows that with the increase of laser power, the effect of interlayer cooling will become more and more significant.

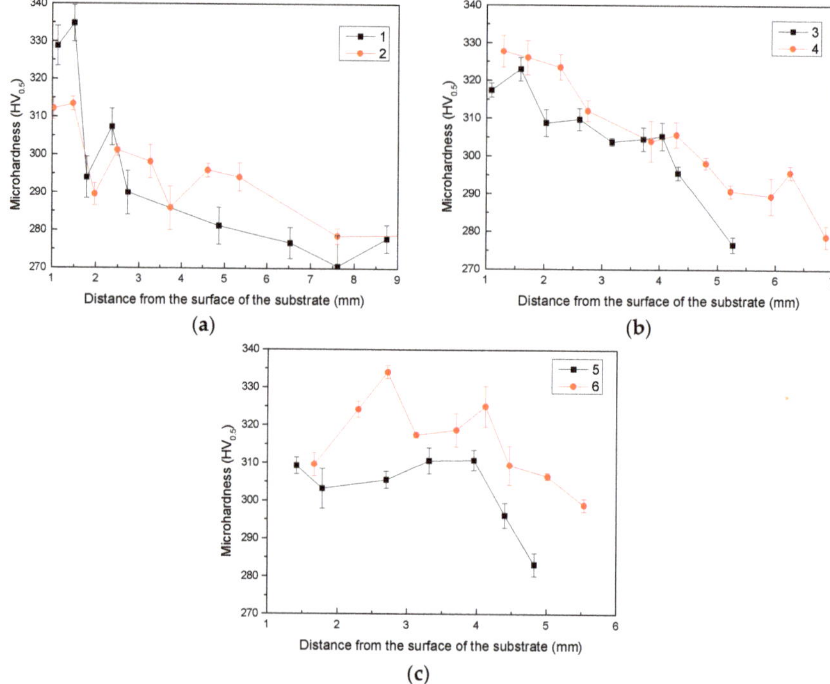

Figure 12. Microhardness (HV) distribution of 6 groups of the single track along the depth direction from the top surface of the substrate: (**a**) Sample 1 and 2; (**b**) Sample 3 and 4; (**c**) Sample 5 and 6.

4. Conclusions

In metal additive manufacturing of non-interlayer cooling deposition, the macro-size is difficult to control due to the heat accumulation effect. In this work, the interlayer cooling effect was studied by cladding Ni-based coatings on a ductile iron substrate with different powers. By comparing the microstructures and mechanical properties of the Ni-based alloy cladding on ductile iron with and without interlayer cooling, the following conclusions can be drawn:

- With the development of the cladding process, due to the heat storage effect and remelting, the increase in thickness becomes smaller and smaller. Thus, it becomes more difficult to control the macro-size. However, through the interlayer cooling, the thickness increment between layers will be more uniform, and the dilution rate will be smaller.
- Compared with the non-interlayer cooling deposition, due to better heat dissipation, the interlayer cooling deposited specimens have less impurities and more uniform microstructure. Meanwhile, the average grain size is refined, and the dendrite growth is inhibited. In addition, the difference between the two grain sizes will become more and more significant with the increase of the power. At the same time, the increase of the power makes the grain sizes of non-interlaminar cooling samples increase significantly, while the grain sizes of interlayer cooling samples have no significant change. It plays an important role in improving the use of the power and has a positive impact on its corresponding mechanical properties.
- Compared with the non-interlayer cooling deposition, the interlayer cooling deposition specimens have higher hardness. Furthermore, as the power increases, the difference between the hardness of the two specimens becomes larger.

Author Contributions: Investigation, Z.Z. and W.H.; Data Curation, J.C. and F.X.; Writing—Original Draft Preparation, Y.H.; Writing—Review and Editing, Y.L. and J.S.; Project Administration, Y.H.; Funding Acquisition, J.Y. and Y.H. All authors have read and agreed to the published version of the manuscript.

Funding: This research was funded by the National Key Research and Development Program of China (Grant No. 2017YFB1103200), Natural Science Foundation of Jiangsu Universities (Grant Nos. 17KJB510031, BK20190712), Six Talent Peaks Project in Jiangsu Province (Grant No. TD-GDZB-009) and Postgraduate Research & Practice Innovation Program of Jiangsu Province (Grant No. KYCX19_0814), Nantong Key Laboratory of 3D printing technology and Application CP12016002.

Conflicts of Interest: The authors declare no conflict of interest.

References

1. Labrecque, C.; Gagné, M. Ductile iron: Fifty years of continuous development. *Can. Metall. Qual.* **1998**, *37*, 343–378.
2. Zeng, D.W.; Xie, C.S.; Yung, K.C. Investigation of laser surface alloying of copper on high nickel austenitic ductile iron. *Mater. Sci. Eng.* **2003**, *333*, 23–231. [CrossRef]
3. Li, R.F.; Qiu, Y.; Zhang, Q.C. Finite element simulation of temperature and stress field for laser cladded nickel-based amorphous composite Coatings. *Coatings* **2018**, *8*, 336. [CrossRef]
4. Huang, Y.J.; Zeng, X. Investigation on cracking behavior of Ni-based coating by laser-induction hybrid cladding. *Appl. Surf. Sci.* **2010**, *256*, 5985–5992. [CrossRef]
5. Zhou, S.; Dai, X.; Zeng, X. Effects of processing parameters on structure of Ni-based WC composite coatings during laser induction hybrid rapid cladding. *Appl. Surf. Sci.* **2009**, *255*, 8494–8500. [CrossRef]
6. Huang, S.W.; Nolan, D.; Brandt, M. Pre-placed WC/Ni clad layers produced with a pulsed Nd: YAG laser via optical fibers. *Surf. Coat. Technol.* **2003**, *165*, 26–34. [CrossRef]
7. Mughal, M.P.; Fawad, H.; Mufti, R. Finite element prediction of thermal stresses and deformations in layered manufacturing of metallic parts. *Acta Mech.* **2006**, *183*, 61–79. [CrossRef]
8. Mughal, M.P.; Fawad, H.; Mufti, R. Numerical thermal analysis to study the effect of static contact angle on the cooling rate of a molten metal droplet. *Numer. Heat Transf. A* **2006**, *49*, 95–107. [CrossRef]
9. Denlinger, E.R.; Heigel, J.C.; Michaleris, P. Effect of inter-layer dwell time on distortion and residual stress in additive manufacturing of titanium and nickel alloys. *J. Mater. Process. Technol.* **2015**, *215*, 123–131. [CrossRef]
10. Chen, Y.; Lu, F.; Zhang, K. Dendritic microstructure and hot cracking of laser additive manufactured Inconel 718 under improved base cooling. *J. Alloys Compd.* **2016**, *670*, 312–321. [CrossRef]
11. Weng, F.; Yu, H.J.; Chen, C.Z. Effect of process parameters on the microstructure evolution and wear property of the laser cladding coatings on Ti–6Al–4V alloy. *J. Alloys Compd.* **2017**, *692*, 989–996. [CrossRef]
12. Liu, F.C.; Lin, X.; Yang, G.L. Microstructures and mechanical properties of laser solid formed nickel base superalloy Inconel 718 prepared in different atmospheres. *J. Metals* **2010**, *46*, 1047–1054.
13. Cheng, Y.H.; Cui, R.; Wang, H.Z. Effect of processing parameters of laser on microstructure and properties of cladding 42CrMo steel. *Int. J. Adv. Des. Manuf. Technol.* **2018**, *96*, 1715–1724. [CrossRef]
14. Radhakrishnan, B.; Thompson, R.G. Solidification of the nickel-base superalloy 718: A phase diagram approach. *Metall. Trans. A* **1989**, *20*, 2866–2868. [CrossRef]
15. Spencer, J.; Dickens, P.; Wykeo, C. Rapid prototyping of metal parts by 3D welding. *Inst. Mech. Eng. B J. Eng. Manuf.* **1998**, *212*, 175–181. [CrossRef]
16. Song, W.L.; Zhu, P.D.; Cui, K. Effect of Ni content on cracking susceptibility and microstructure of laser-clad Fe–Cr–Ni–B–Si alloy. *Surf. Coat. Technol.* **1996**, *80*, 279–282.
17. Qian, M.; Lim, I.C.; Chen, Z.D. Parametric studies of laser cladding processes. *J. Mater. Process. Technol.* **1997**, *63*, 590–593. [CrossRef]
18. Liu, H.X.; Wang, C.Q.; Zhang, X.W. Improving the corrosion resistance and mechanical property of 45 steel surface by laser cladding with Ni60CuMoW alloy powder. *Surf. Coat. Technol.* **2013**, *228*, S296–S300. [CrossRef]
19. Ke, L.; Zhu, H.; Yin, J. Effects of peak laser power on laser micro sintering of nickel powder by pulsed Nd:YAG laser. *Rapid Prototyp. J.* **2014**, *20*, 328–335. [CrossRef]
20. Singh, A.; Dahotre, N.B. Laser in situ synthesis of mixed carbide coating on steel. *Mater. Sci.* **2004**, *39*, 4553–4560. [CrossRef]

21. Lima, M.S.F.; Ladario, F.P.; Riva, R. Microstructural analyses of the nanoparticles obtained after laser irradiation of Ti and W in ethanol. *Appl. Surf. Sci.* **2006**, *252*, 4420–4424. [CrossRef]
22. Gao, W.; Zhang, Z.; Zhao, S. Effect of a small addition of Ti on the Fe-based coating by laser cladding. *Surf. Coat. Technol.* **2016**, *291*, 423–429. [CrossRef]

Article

Tribocorrosion Behavior of Inconel 718 Fabricated by Laser Powder Bed Fusion-Based Additive Manufacturing

Arpith Siddaiah [1], Ashish Kasar [1], Pankaj Kumar [2,3], Javed Akram [4], Manoranjan Misra [2] and Pradeep L. Menezes [1,*]

1 Department of Mechanical Engineering, University of Nevada, Reno, NV 89557, USA; asiddaiah@nevada.unr.edu (A.S.); akasar@nevada.unr.edu (A.K.)
2 Department of Chemical and Materials Engineering, University of Nevada, Reno, NV 89557, USA; pankaj@unm.edu (P.K.); misra@unr.edu (M.M.)
3 Department of Mechanical Engineering, University of New Mexico, Albuquerque, NM 87131, USA
4 Ansys, 1794 Olympic Parkway, Site#110, Park City, UT 84098, USA; jaavedakram@gmail.com
* Correspondence: pmenezes@unr.edu; Tel.: +1-(775)-682-7413

Abstract: Additive manufacturing (AM) by laser powder bed fusion (LPBF) has gained significant research attention to fabricate complex 3D Inconel alloy components for jet engines. The strategic advantages of LPBF-based AM to fabricate jet components for aerospace applications are well reported. The jet components are exposed to a high degree of vibration during the jet operation in a variable aqueous environment. The combined vibration and the aqueous environment create a tribological condition that can accelerate the failure mechanism. Therefore, it is critical to understand the tribocorrosion behavior of the Inconel alloy. In the present work, tribocorrosion behavior of the LPBF fabricated standalone coating of Inconel 718 in the 3.5% NaCl aqueous solution is presented. The LPBF fabricated samples are analyzed to determine the impact of porosity, generated as a result of LPBF, on the triobocorrosion behavior of AM Inconel 718. The study includes potentiodynamic tests, cathodic polarization, along with OCP measurements. The corrosive environment is found to increase the wear by 29.24% and 49.5% without the initiation of corrosion in the case of AM and wrought Inconel 718, respectively. A corrosion accelerated wear form of tribocorrosion is observed for Inconel 718. Additionally, the corrosive environment has a significant effect on wear even when the Inconel 718 surface is in equilibrium potential with the corrosive environment and no corrosion potential scan is applied. This study provides an insight into a critical aspect of the AM Inconel components.

Keywords: additive manufacturing; Inconel 718; friction; wear; tribocorrosion; corrosion

Citation: Siddaiah, A.; Kasar, A.; Kumar, P.; Akram, J.; Misra, M.; Menezes, P.L. Tribocorrosion Behavior of Inconel 718 Fabricated by Laser Powder Bed Fusion-Based Additive Manufacturing. *Coatings* **2021**, *11*, 195. https://doi.org/10.3390/coatings11020195

Academic Editor: Kevin Plucknett
Received: 21 December 2020
Accepted: 5 February 2021
Published: 8 February 2021

1. Introduction

Inconel 718 is a super-alloy. Due to its excellent high-temperature performance, it is widely used in aircraft turbojet engines as discs, blades, and casing for high-pressure sections of aircraft engine components. It also finds application in rocket engines and cryogenic environments due to its good toughness at low temperatures, which protects the components from brittle fracture. Due to excellent mechanical properties and wear resistance, the Inconel 718 is favorable for aerospace, marine and chemical industries. However, the higher shear strength and low material removal rate of Inconel 718 create difficulty in machining, especially for complex parts where tight dimensional accuracy is required [1–3]. Additive manufacturing (AM) provides an excellent opportunity to overcome the difficulty of conventional machining to develop a complex structure of Inconel 718 [4,5].

Among the different methods of AM techniques, laser powder bed fusion (LPBF) has gained tremendous attention due to its capability to fabricate a fully dense component [6,7]. LPBF is also referred as direct laser sintering or selective laser melting. LPBF technique involves layer-by-layer formation where a layer of powder is applied to the platform/substrate and subsequently melted using laser. The melting of the powders causes bonding between powders and forms the first layer. Similarly, successive layers of powders are formed to fabricate the coating/parts. LPBF technique has been used to fabricate different standalone coatings (ref.) and 3D components (ref.) of various materials such as Ti6Al4V coating [8], tantalum coating [9], Ti-SiC coating [10], 316L stainless steel [11], and Inconel 718 [12] etc.

LPBF has enabled customization of Inconel 718 components with easy production of complex shapes and geometries. During its operation, especially in aircraft and jet applications, the material may come in contact with various extreme and aqueous environments in addition to tribo-contacts with other materials [13]. These conditions can cause early failure or reduced performance of AM Inconel 718 components if the AM design or fabrication process has any flaws. It is also well known that the Inconel 718 provides excellent corrosion resistance. It has been shown that the AM Inconel 718 can maintain pitting corrosion resistance even after low-temperature solution treatments, while high-temperature solution treatments can cause some weakening of the pitting corrosion resistance [13–15]. Additionally, the electrochemical polishing process can significantly improve pitting corrosion resistance. Build direction using LPBF has also shown a significant influence on microstructure and corrosion performance of Inconel 718. The corrosion resistance of the AM Inconel 718 samples increases with the increasing incline angle of LPBF process, and this can be rationalized by considering changes to the grain boundary area [16]. To make the AM Inconel 718 tribo-contact interfaces functionally resistant in extreme environmental conditions, it is necessary to investigate the wear along with the electrochemical behavior of these alloys. Hence, understanding the tribological and tribocorrosion behavior of functionally processed Inconel 718 materials are of practical and theoretical importance.

2. Materials and Methods

2.1. Materials and Electrolyte

In this investigation, 3D-printed Inconel 718 samples of size 10 $\times$ 10 $\times$ 5 mm^3 using the LPBF process were manufactured using SLM^R 500 machine (SLM Solutions, Wixom, MI, USA) of maximum power 400 W. The deposition parameters for the sample preparation were: Laser power: 165 W, scan speed: 800 mm/s, and layer thickness: 0.4 mm. Detailed manufacturing details are given elsewhere [7]. The porosity of <1% in the LPBF-processed sample has been observed. The microstructure of the as-fabricated cube sample showed large and interconnected directional columnar grains with random columnar grain as discussed in detail in [7]. A wrought Inconel 718 (purchased from McMaster Carr, Elmhurst, IL, USA) has been used as a control material for comparison of the properties being investigated. The chemical composition of both wrought and as LPBF samples qualifies as ASTM B637 for Inconel 718 alloy [17]. The hardness measurement was carried out using a nano-indenter (T1950 Triboindenter, Hysitron, Minneapolis, MN, USA) equipped with a Berkovich indenter. The nanoindentations were performed at different locations using a maximum force of 5 mN. The minimum distance between two indentations was 100 μm. The hardness value of AM Inconel is 5.36 $\pm$ 0.30 GPa, which is slightly higher than the hardness value of Wrought Inconel (4.93 $\pm$ 0.27 GPa).

The tribocorrosion studies were performed on as LPBF and as purchased wrought samples without any further heat treatment. For this study, square plate samples of Inconel 718 (both the LPBF and wrought) with a 1-inch width and 0.5-inch thickness were used to evaluate its tribocorrosion behavior. The surface of the samples was polished using SiC sandpapers with grit sizes starting from 320 (~46 μm abrasive particles) up to 1200 (~15 μm abrasive particles), followed by fine polishing using three-micron diamond suspension until a surface roughness conditions of R_a = 0.1 μm was achieved. The counter material for the tribocorrosion testing was a non-conductive ceramic material, alumina ball of 6.35 mm diameter. The present tribocorrosion study simulates a corrosive lubricated environment according to ASTM B895-99 [18], where 3.5% NaCl solution is used as the electrolyte/lubricant. The test parameters in the present study concentrated on analyzing the tribocorrosion behavior in laboratory conditions. The real applications of AM Inconel 718, such as those in aircrafts may have much more complex boundary conditions, which will include, but may not be limited to, fretting and other vibrations, reciprocating sliding, and other such degradation effects.

2.2. Open Circuit Potential (OCP) Measurements

The OCP measurements were carried out on wrought and AM Inconel 718 surfaces to understand and compare their open circuit kinetics in 3.5% NaCl. The OCP, also known as the corrosion potential is a mixed potential that depends on the rate of the anodic as well as the cathodic reactions. If the corrosion cell involves one dissolution reaction and one cathodic reaction, the corrosion potential will be between the reversible potentials of the two half-reactions. Understanding this evolution of OCP will provide an insight into the kinetics of corrosion without wear. The experiments were conducted using three-electrode configuration, which consisted of Inconel 718 surface as the working electrode, standard calomel electrode (SCE) as the reference electrode, and 99% pure graphite as the counter electrode. Figure 1 schematically illustrates the tribocorrosion test setup used for the experiments. The OCP was measured using a Gamry reference 3000 potentiostat. All potential in the present study is reported with reference to SCE, which has a potential of 0.250 V vs. SHE (standard hydrogen electrode) at 25 °C. A working electrode area of 2.5 cm^2 was exposed to the electrolyte. Each test was carried out three times to ensure repeatability of results.

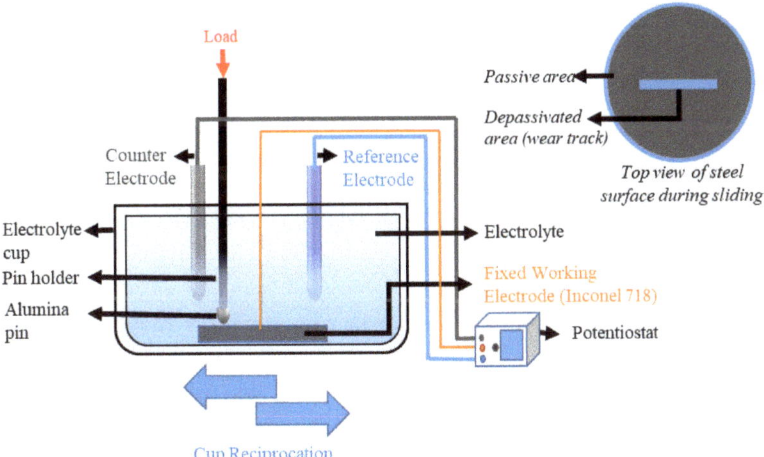

Figure 1. Tribocorrosion test setup and view of the Inconel 718 sample/working electrode.

2.3. Tribocorrosion Tests

The synergistic effect between wear and corrosion (S) affects the total material loss rate (T) as

$$T = W_o + C_o + S \tag{1}$$

where W_o is the wear component in the absence of corrosion and C_o is the corrosion component in the absence of wear. The synergistic component S can be written as

$$S = \Delta W_c + \Delta C_w \tag{2}$$

where ΔW_c is the change in wear rate due to corrosion, and ΔC_w is the change in corrosion rate due to wear. The total wear component is $W_o + \Delta W_c$, and the total corrosion component is $C_o + \Delta C_w$. These components in a tribocorrosive system will be measured using a combination tribological and electrochemical experiments. The corrosion rate of a material can be obtained using Faraday's law:

$$V_{ch} = QM/nF\rho \quad (3)$$

where V_{ch} is the volume of metal transformed by anodic oxidation, $Q = \int Idt$ is the electric charge passed over the time of the experiment, M is the atomic mass of the metal, n is the charge number for the oxidation reaction, F is the Faraday constant, and ρ is the density of the metal. Whereas the wear components will be quantified using the Archard's law [19], the relationship between the hardness of the wear volume can be described as:

$$V/L = kN/H \quad (4)$$

where V is the volume of wear, L is the sliding distance, N is the normal load, H is the hardness of the softer one in the two contacting materials, and k is the wear coefficient, which is valid to both the adhesive and abrasive wears.

Figure 1 schematically illustrates the tribocorrosion test setup used for the experiments. To isolate wear-corrosion synergism in a standard seawater condition of 3.5%wt NaCl, the following series of experiments were performed as per the ASTM G119 [20] for wrought and AM Inconel 718 disks with surface roughness conditions of R_a = 0.1 µm. The counter/ball material used will be alumina of 6.35 mm diameter.

(1) First, the experiment will be performed using the tribocorrosion monitoring test setup (Figure 1) developed in-house [21]. The setup consists of a pin-on-disk setup, and an electrochemical cell with three-electrode configuration. The total material wear rate (T) of the disk was be measured while monitoring the equilibrium potential during sliding against alumina ball.

(2) Second, the rate of degradation of the disk was determined by measuring corrosion rate in the absence of wear (C_o). The same setup (Figure 1) was used without sliding to measure the corrosion rate without wear. The corrosion rate was calculated by a series of electrochemical measurements involving, (i) measuring the equilibrium potential of the disk material, (ii) polarization resistance measurements, (iii) measuring the equilibrium potential of the disk material, and (iv) potentiodynamic polarization tests. The electrochemical corrosion current density was calculated using the results from these tests, and the C_o will be obtained by applying Faraday's law (Equation (3)).

(3) Third, the rate of disk degradation in the absence of corrosion (W_o) was determined. The working electrode on the test setup was polarized cathodic with respect to corrosion potential (E_{corr}) to suppress corrosion, and the wear data was collected. The rate of material loss on the disk at the end of the experiment was calculated in Equation (4). This rate of material loss will provide the W_o.

(4) Fourth, the total corrosion component (C_w) was estimated. Experiments were performed using the test set-up (Figure 1). This stage involved a wear test in addition to the electrochemical experiments from the second stage. The electrochemical corrosion current density as influenced by the wear was calculated using the results from these tests. The C_w was then obtained by applying Faraday's law (Equation (3)).

3. Results and Discussion

In the present study, 2025.8 mm^2 surface area of Inconel 718 was in contact with the electrolyte throughout the tribocorrosion testing. Within this area of contact, the cyclic 10 mm reciprocating sliding of alumina ball on Inconel 718 surface was the depassivated area (anode), and the remaining unworn surface was the passive area (cathode), thus completing the galvanic couple. The tribocorrosion test was conducted at ±5 mm/stroke sliding cycle for 60 min; the sliding speed was 0.01 mm/s at a normal load of 50 N. The tests were conducted at laboratory conditions of 24 °C lab temperature and 20% relative humidity. During the reciprocating tests, potential and current values were recorded on the potentiostat while the wear and friction force was recorded on the tribometer, in-situ. A summary of the observed and calculated (using Equations (1)–(4)) results are shown in Table 1, and their implications are discussed in the following sections. A strong synergistic effect was observed on the Inconel Wrought sample with a synergistic factor of 124.93. It is significantly higher compared to that of the Inconel AM sample. By resolving the synergistic effect into wear and corrosion, the component shows a strong effect of corrosion on wear for Inconel wrought. The higher value of ΔW_c in the case of Inconel AM suggests that mechanical wear is drastically influenced by corrosion. Whereas the effect of wear on corrosion is negative that is possibly due to corrosion protection by wear debris. These synergistic terms are discussed in terms of wear volume and corrosion rate in the next subsection.

Table 1. Material loss rate for Inconel AM and Wrought sample during tribocorrosion.

Sample	Material Loss Rate, ($mm^3/mm^2 \cdot year$)						
	T	W_o	C_o	C_w	S	ΔC_w	ΔW_c
Inconel AM	107.05	106.26	0.08	0.06	0.07	−0.02	0.72
Inconel Wrought	225.40	100.40	0.07	0.06	124.93	−0.01	124.94

3.1. Wear Volume during Tribocorrosion

The wear behavior was measured for wrought and AM Inconel 718 during the OCP, potentiodynamic, and cathodic polarization tests. The wear volumes with respective wear track profile results for the wrought sample are presented in Figure 2 and for the AM sample are presented in Figure 3. The wear tracks were profiled and analyzed using a 3D optical profilometer. The 3D profiles were used to accurately extract the wear volume under each test condition and were compared. It can be observed in Figure 2 that the wrought Inconel 718 undergoes wear of 2.57 mm^3 under the influence of corrosion and wear during the potentiodynamic test. The wear under OCP conditions in the sample is in equilibrium potential with the corrosive environment and is found to be 2.28 mm^3. Under cathodic polarization conditions, the effects of corrosion have been electrochemically removed to isolate the wear under the given corrosive environment; the wear is found to be 1.15 mm^3.

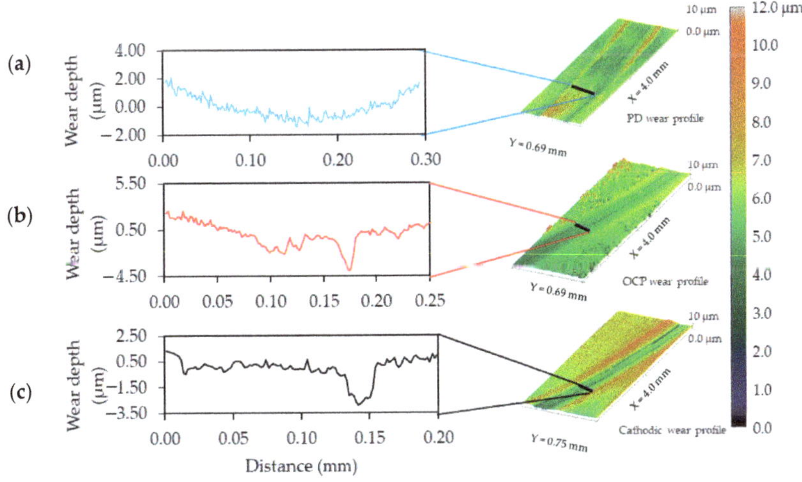

Figure 2. Wear on wrought Inconel 718 after (**a**) potentiodynamic test, (**b**) OCP condition, and (**c**) cathodic polarization test shown alongside a common color scale bar.

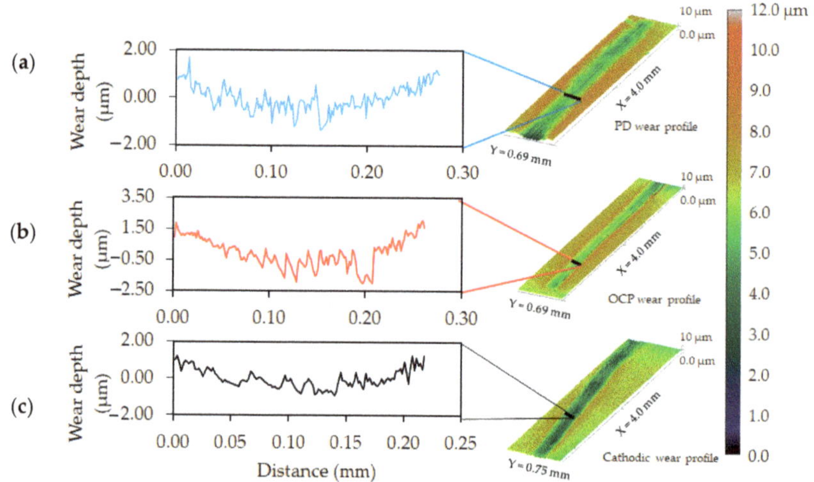

Figure 3. Wear on AM Inconel 718 after (**a**) potentiodynamic test, (**b**) OCP condition, and (**c**) cathodic polarization test shown alongside a common color scale bar.

Comparing the potentiodynamic and OCP wear volume results, it is observed that the wear is accelerated due to the onset of corrosion by a factor of 11.28% in the corrosive environment. It is believed that this is due to the small roughness reduction initially at the onset of corrosion, exposing the corrosion surface area and oxide film defect, increasing the susceptibility to corrosion product formation [22]. The corrosive environment itself seems to have a significant influence on surface degradation, as when the OCP and cathodic polarization wear volumes are compared, the wear seems to increase by 49.5%. The synergistic effect considering the corrosion accelerated wear in the corrosive environment the wear volume increased by 55.25%. This shows the drastic effect of aqueous corrosive environment on wrought Inconel 718.

In Figure 3, it can be observed that for AM Inconel 718, the potentiodyanamic wear volume is 1.22 mm^3, the OCP wear volume is 1.71 mm^3, and the cathodic polarization wear volume is 1.21 mm^3. It is evident from comparing the OCP and cathodic polarization wear conditions that the corrosive environment increases the wear by 29.24% without the initiation of corrosion. It is interesting to note from the potentiodynamic polarization and OCP wear volumes that the initiation of corrosion actually decreases wear volume by 28.7%. This indicates that the AM Inconel 718 has a directional solidification during its fabrication and can be equal to or exceed the mechanical properties of wrought counterpart, in this case it is observed as reduced wear volume. Studies have shown that the corrosion resistance of the AM Inconel 718 samples increases with the increasing incline angle of LPBF process, and this can be rationalized by considering changes to the grain boundary area [16]. The synergistic effect considering the corrosion accelerated wear is found be extremely low at only ~1% increase in wear volume. It is expected that there is a small degree of electropolishing that occurs on the surface of the AM Inconel 718 in tribocorrosive environment, which can increase pitting corrosion resistance [23]. These results indicate that AM of Inconel 718 has a significant influence on the tribocorrosion performance as compared to the wrought counterpart. The corrosion results are further investigated in the following section.

3.2. Corrosion Rate during Tribocorrosion

The corrosion rate was calculated using the corrosion current (I_{corr}) obtained during the tribocorrosion tests for wrought and AM Inconel 718. The corrosion rates were calculated for both with and without wear conditions as described in the tribocorrosion test methodology. The observed Tafel plot, along with their respective corrosion rates, is shown in Figure 4. The shift in corrosion potential and corrosion current density due to the effect of corrosion and tribocorrosion can be clearly observed in this figure.

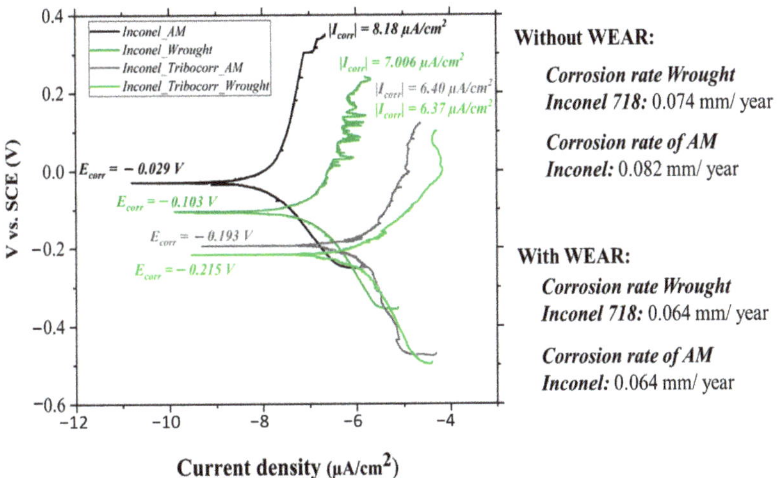

Figure 4. Tafel plots and corrosion rate of wrought and AM Inconel 718 for various tribocorrosion test conditions.

It can be observed in Figure 4 that the corrosion potential (E_{corr}) and corrosion current (I_{corr}) under without wear condition for AM Inconel 718 is −0.029 V and 8.18 μA/cm^2 as compared to −0.103 V and 7.006 μA/cm^2 for wrought Inconel 718, respectively. This indicates that the additively manufacturing Inconel 718 provides an enhanced corrosion resistance potential, which is the point of corrosion onset. This resistance potential for AM Inconel 718 is found to be 71.8% better than that for the wrought counterpart. Even though the corrosion rate of the under the same condition of without wear is found to be slightly more for AM Inconel 718 at 0.082 mm/year as compared to the wrought Inconel 718 at 0.074 mm/year, it can be expected that the onset of corrosion is delayed by 71.8% for AM Inconel 718. Hence, the observed reduction in wear under OCP conditions for AM Inconel 718 as compared to the wrought counterpart (Figures 2b and 3b, respectively), where no corrosion potential was applied. Under the OCP wear condition, the Inconel 718 showed wear volumes of 1.71 mm^3 and 2.28 mm^3 for AM and wrought materials, respectively. The slightly lower corrosion current during the absence of wear for wrought Inconel 718 is found to be what provides it a 9.8% better corrosion resistance without wear but with an early onset of corrosion as compared to the AM Inconel 718.

Considering the wear with corrosion (tribocorrosion) it is observed that the E_{corr} and I_{corr} for both AM and wrought Inconel 718 shift to have an overall reduced corrosion rate, which confirms that the observed surface degradation is a corrosion accelerated wear. The wear volume is observed to increase in the same conditions (Figures 2a and 3a), but due to the delayed onset of corrosion, the wear in the case of AM Inconel 718 (~1% increase) is minimal as compared to wrought Inconel 718 (55.25% increase). This unique behavior of Inconel 718 can be attributed to the anodic-to-cathodic transition potential (reverse portion of the potentidynamic scan) is more noble than corrosion potential. It can be believed, according to the Silverman's interpretation [24], that any film created on the reference Inconel 718 surface at the corrosion potential might not be very passivating and would allow uniform corrosion to occur at the corrosion potential. The above analysis can be summarized (in a similar way as in Davydov et al. [25]) by stating that using AM manufacturing significantly increases resistance to the pitting corrosion of Inconel 718, but does not increase its ability for passivation in NaCl solution.

4. Conclusions

The wear-corrosion synergism for the tribo-pair was be calculated using Equations (1) and (2). A positive synergy between wear and corrosion is highly undesirable. The experiment as detailed by ASTM G119 was used to analyze the tribocorrosion behavior of Inconel 718. The described method of tribocorrosion testing enables the measurement of the galvanic current between the passive (cathode, unworn area) and the depassivated (anode, worn area) areas of the sample surface at OCP. The following conclusions can be drawn from the present study:

(a) The corrosive environment increases the wear by 29.24% and 49.5% without the initiation of corrosion in the case of AM and wrought Inconel 718, respectively.
(b) The onset of corrosion in case of AM Inconel 718 is delayed by 71.8% as compared to the wrought counterpart.
(c) The late onset of corrosion degradation (high corrosion resistance potential) provides better tribocorrosion resistance in the case of AM Inconel 718.
(d) Both AM and wrought Inconel 718 have similar corrosion rates during tribocorrosion, but the wear volume in case of AM is much lower (1.22 mm^3) as compared to wrought (2.57 mm^3)
(e) A corrosion-accelerated wear form of tribocorrosion is observed for Inconel 718.
(f) The corrosive environment has a significant effect on wear even when the Inconel 718 surface is in equilibrium potential with the corrosive environment, and no corrosion potential scan is applied.

Author Contributions: Conceptualization: A.S., P.K., and P.L.M.; Methodology: A.S. and A.K.; Software, A.S. and A.K.; Validation: A.S. and A.K.; Formal analysis: A.S., A.K., P.K., M.M., and P.L.M.; Investigation, A.S., A.K., and P.L.M.; Resources: J.A., P.K., M.M., and P.L.M.; Data curation, A.S. and A.K.; Writing—original draft preparation, A.S.; Writing—review and editing: A.S., A.K., P.K., J.A., and P.L.M.; Visualization: A.S. and A.K.; Supervision: P.K., M.M., and P.L.M.; Project administration: P.K. and P.L.M. All authors have read and agreed to the published version of the manuscript.

Funding: This research received no external funding.

Institutional Review Board Statement: Not applicable.

Informed Consent Statement: Not applicable.

Data Availability Statement: Data available on request due to restrictions e.g., privacy or ethical. The data presented in this study are available on request from the corresponding author. The data are not publicly available due to propriority materials, equipment, and fabrication process being studied.

Acknowledgments: The authors appreciate the facility and financial support by startup funding from the Department of Mechanical Engineering, and the Department of Chemical and Materials Engineering at the University of Nevada, Reno.

Conflicts of Interest: The authors declare no conflict of interest.

References

1. Rahman, M.; Seah, W.K.H.; Teo, T.T. The machinability of inconel 718. *J. Mater. Process. Technol.* **1997**, *63*, 199–204. [CrossRef]
2. Xavior, M.A.; Patil, M.; Maiti, A.; Raj, M.; Lohia, N. Machinability studies on INCONEL 718. IOP Conference Series: Materials Science and Engineering. In Proceedings of the International Conference on Advances in Materials and Manufacturing Applications (IConAMMA-2016), Bangalore, India, 14–16 July 2016.
3. Iturbe, A.; Giraud, E.; Hormaetxe, E.; Garay, A.; Germain, G.; Ostolaza, K.; Arrazola, P.J. Mechanical characterization and modelling of Inconel 718 material behavior for machining process assessment. *Mater. Sci. Eng. A* **2017**, *682*, 441–453. [CrossRef]
4. Wang, X.; Gong, X.; Chou, K. Review on powder-bed laser additive manufacturing of Inconel 718 parts. *Proc. Inst. Mech. Eng. Part B J. Eng. Manuf.* **2016**, *231*, 1890–1903. [CrossRef]
5. Jia, Q.; Gu, D. Selective laser melting additive manufacturing of Inconel 718 superalloy parts: Densification, microstructure and properties. *J. Alloys Compd.* **2014**, *585*, 713–721. [CrossRef]

6. Zhang, B.; Liao, H.; Coddet, C. Selective laser melting commercially pure Ti under vacuum. *Vacuum* **2013**, *95*, 25–29. [CrossRef]
7. Kumar, P.; Farah, J.; Akram, J.; Teng, C.; Ginn, J.; Misra, M. Influence of laser processing parameters on porosity in Inconel 718 during additive manufacturing. *Int. J. Adv. Manuf. Technol.* **2019**, *103*, 1497–1507. [CrossRef]
8. Morton, P.A.; Taylor, H.C.; Murr, L.E.; Delgado, O.G.; Terrazas, C.A.; Wicker, R.B. In situ selective laser gas nitriding for composite TiN/Ti-6Al-4V fabrication via laser powder bed fusion. *J. Mater. Sci. Technol.* **2020**, *45*, 98–107. [CrossRef]
9. Fox, P.; Pogson, S.; Sutcliffe, C.J.; Jones, E. Interface interactions between porous titanium/tantalum coatings, produced by Selective Laser Melting (SLM), on a cobalt–chromium alloy. *Surf. Coat. Technol.* **2008**, *202*, 5001–5007. [CrossRef]
10. Krakhmalev, P.; Yadroitsev, I. Microstructure and properties of intermetallic composite coatings fabricated by selective laser melting of Ti–SiC powder mixtures. *Intermetallics* **2014**, *46*, 147–155. [CrossRef]
11. Heiden, M.J.; Deibler, L.A.; Rodelas, J.M.; Koepke, J.R.; Tung, D.J.; Saiz, D.J.; Jared, B.H. Evolution of 316L stainless steel feedstock due to laser powder bed fusion process. *Addit. Manuf.* **2019**, *25*, 84–103. [CrossRef]
12. Barros, R.; Silva, F.J.G.; Gouveia, R.M.; Saboori, A.; Marchese, G.; Biamino, S.; Salmi, A.; Atzeni, E. Laser powder bed fusion of Inconel 718: Residual stress analysis before and after heat treatment. *Metals* **2019**, *9*, 1290. [CrossRef]
13. Luo, S.; Huang, W.; Yang, H.; Yang, J.; Wang, Z.; Zeng, X. Microstructural evolution and corrosion behaviors of Inconel 718 alloy produced by selective laser melting following different heat treatments. *Addit. Manuf.* **2019**, *30*, 100875. [CrossRef]
14. Zhang, B.; Xiu, M.; Tan, Y.T.; Wei, J.; Wang, P. Pitting corrosion of SLM Inconel 718 sample under surface and heat treatments. *Appl. Surf. Sci.* **2019**, *490*, 556–567. [CrossRef]
15. Zhang, L.N.; Ojo, O.A. Corrosion behavior of wire arc additive manufactured Inconel 718 superalloy. *J. Alloys Compd.* **2020**, *829*, 154455. [CrossRef]
16. Du, D.; Dong, A.; Shu, D.; Zhu, G.; Sun, B.; Li, X.; Lavernia, E. Influence of build orientation on microstructure, mechanical and corrosion behavior of Inconel 718 processed by selective laser melting. *Mater. Sci. Eng. A* **2019**, *760*, 469–480. [CrossRef]
17. ASTM B637-18. *Standard Specification for Precipitation-Hardening and Cold Worked Nickel Alloy Bars, Forgings, and Forging Stock for Moderate or High Temperature Service*; ASTM International: West Conshohocken, PA, USA, 2018.
18. ASTM B895-16(2020)e1. *Standard Test Methods for Evaluating the Corrosion Resistance of Stainless Steel Powder Metallurgy (PM) Parts/Specimens by Immersion in a Sodium Chloride Solution*; ASTM International: West Conshohocken, PA, USA, 2020.
19. Archard, J.F. Contact and Rubbing of Flat Surfaces. *J. Appl. Phys.* **1953**, *24*, 981–988. [CrossRef]
20. ASTM G119-09(2016). *Standard Guide for Determining Synergism Between Wear and Corrosion*; ASTM International: West Conshohocken, PA, USA, 2016.
21. Siddaiah, A.; Khan, A.Z.; Ramachandran, R.; Menezes, L.P. Performance Analysis of Retrofitted Tribo-Corrosion Test Rig for Monitoring In Situ Oil Conditions. *Materials* **2017**, *10*, 1145. [CrossRef] [PubMed]
22. Wang, J.; Xu, J.; Zhang, X.; Ren, X.; Song, X.; Chen, X. An investigation of surface corrosion behavior of Inconel 718 after robotic belt grinding. *Materials* **2018**, *11*, 2440. [CrossRef] [PubMed]
23. Guo, P.; Lin, X.; Li, J.; Zhang, Y.; Song, M.; Huang, W. Electrochemical behavior of Inconel 718 fabricated by laser solid forming on different sections. *Corros. Sci.* **2018**, *132*, 79–89. [CrossRef]
24. Silverman, D.C. Tutorial on Cyclic Potentiodynamic Polarization Technique. In Proceedings of the CORROSION/98 Research Topical Symposium; NACE: San Diego, CA, USA, 1998; p. 21.
25. Davydov, A.D.; Shaldaeva, V.S.; Malofeevaa, A.N.; Chernyshovab, O.V.; Volgina, V.M. Determination of corrosion rate of rhenium and its alloys. *Chem. Eng.* **2014**, *41*, 289–294.

MDPI
St. Alban-Anlage 66
4052 Basel
Switzerland
Tel. +41 61 683 77 34
Fax +41 61 302 89 18
www.mdpi.com

Coatings Editorial Office
E-mail: coatings@mdpi.com
www.mdpi.com/journal/coatings

www.ingramcontent.com/pod-product-compliance
Lightning Source LLC
LaVergne TN
LVHW070617170726
843515LV00004B/112

* 9 7 8 3 0 3 6 5 2 4 8 8 7 *